KOREAN STUDIES SERIES NO. 48

Sasang Constitutional Typology:

A Beginner's Guide to Sasang Constitutional Medicine

Written by Lee Eui-ju, Lee Ji-won, and Yoo Jung-hee

Jimoondang
Seoul

Jimoondang
85 Gwanginsa-gil, Paju-si, Gyeonggi-do, 413-756, Korea
82 Donhwamun-ro, Jongno-gu, Seoul, 110-360, Korea
227 Suttons Lane, Edison, NJ 08817, USA
Phone: 82-2-743-3096 E-mail: edit@jimoon.co.kr
82-2-743-3192~3 E-mail: sale@jimoon.co.kr
Fax: 82-2-743-0227, 82-2-742-4657
Homepage: www.jimoon.co.kr

The National Library of Korea Cataloging-in-Publication (CIP)
Sasang Constitutional Typology: A Beginner's Guide to Sasang Constitutional Medicine
written by Lee Eui-ju, Lee Ji-won, and Yoo Jung-hee
Paju: Jimoondang, 2014 (Korean studies series; No. 48)
ISBN 978-89-6297-152-1 94510
978-89-6297-002-9 (set) 519.74-KDC5 616.075-DDC21 CIP2014008378

Printed in Korea

Preface

More than hundred years have passed since Lee Je-ma's death in 1900. In Korea, the 'O-ring test'[1] that had become popular in the 1990's and the recent hit TV show "Taeyangin Lee Je-ma" have put Constitutional Medicine in the limelight.

In the past 10 years, people have begun to recognize "constitution" in context of the Sasang Constitutional Medicine, a division of Korean Medicine.

Most people who are first interested in Constitutional Medicine want to know what type of constitution they have and how to live a healthy life most compatible with their constitutional type. Then, they begin to learn more about constitutional typology and try to guess what their constitutional type could be. Some people become dubious, which is natural considering that there are not many objective methods for diagnosing the constitutional type. Traditional practitioners had strived in the past to discover an objective tool for constitutional diagnosis, producing a few possibilities such as morphological analysis, blood analysis, and genetical analysis. We are anticipating that more objective ways of diagnosing the constitutional type will be discovered in the near future.

---

1 O-ring test

The original nomenclature is the 'Omura test', named after its discoverer. Since we presently don't know about all the functions of the brain, a test based on the cerebral function will also have certain shortcomings. But the O-ring test has helped us pave the way for measuring 'Italicize' by measuring the strength of a muscle when the person is in touch with a certain object.

It is disturbing to see people consider Sasang Constitutional Medicine as some sort of food remedy with a level of sophistication that simply states "beef is compatible for my constitutional type, but pork is not." Of course, the Sasang Constitutional Medicine does provide information on the compatibility of specific foods for the different constitutional types, which is after all an essential element in preventing and managing pathological conditions. However, there is much more than this to Lee Je-ma's Constitutional Medicine.

The 100th year memorial ceremony for Lee Je-ma

Throughout his life, Lee Je-ma constantly asked questions like "What is the human being?" "What is the difference between a sage and a commoner?" "Why does the pathological development of the same disease differ among people?" "Why are certain people more likely to fall prey to particular diseases while others are immune to them?" He studied Confucianism and Korean Medicine and searched for the answers to these questions all his life. In his last days, he compiled the results of his search in his books 『Gyeokchigo (格致藁)』 and 『Dongeui Soose Bowon (東醫壽世保元)』 which established the foundations for Sasang Constitutional Medicine.

The studies made by Lee Je-ma are involved in two main fields: anthropology and Sasang Constitutional Medicine.

The purpose of studying humanity in Lee Je-ma's viewpoint was to pursue the morally accomplished man. But this is rendered almost impossible because of the social environments a person is exposed to, known as "Humanly activity (人事)," which consist of two main elements: "Ah (我)," or a person's internal factors, and "Bi Ah (非我)," or the surrounding external factors. Each constitutional type has its natural state of mind (恒心) and base desires (心慾) which make him morally imperfect. Lee Je-ma proposed his theories on psychology and morality of the humankind and urged people to resolve their "internal" problems by rectifying the self (正己) and to resolve their "external" problems in relationships by understanding the human mind (知人).

The ultimate goal in Sasang Constitutional Medicine is a healthy life. But there are differences in the physical organ structural schemes between the constitutional types which give rise to the differences in physiological and pathological conditions for different constitutional types. Lee Je-ma made profound studies on the pathogenic factor, pathological development, and prevention and treatment of disease that differ for each constitutional type.

Lee Je-ma explains his theories in anthropology and constitutional medicine in a simple and clear way so that even a layman would be able to apply them in practical life. Lee Je-ma mentions in his book 『Dongeui Soose Bowon (東醫壽世保元)』 that "If a town of ten thousand has only one potter, there would be a shortage of pottery, and if a village of a hundred has only one doctor, there would be a shortage of medical care. The art of medicine should be studied and extended, so that every household would have access to medicine and each person would know of his own disease. Only then will there be longevity in the world and preservation of the people." This reflects Lee Je-ma's ideas on the ideal world where people can maintain a healthy life by understanding the mind and body and be able to take care of personal health.

A century has passed since his death, and his contributions to the society are still substantial and valuable. We would all benefit by studying his anthropological views and the theory of Sasang constitutional medicine which can be applied in our practical lives for promoting a healthier lifestyle. If we can accept with open minds and perceive the profoundness of his ideas, we would be able to share his perspectives on the ideal way of living.

I have written this book in hopes that Sasang Constitutional Medicine could be understood better by the public and used in practical life. This book contains materials from several articles and thesis papers that I had written hoping to introduce and promote Sasang Constitutional Medicine to the public. I humbly admit I am lacking in experience and knowledge, and I cannot help fearing that this book may detract from the profoundness of Lee Je-ma's accomplishments. Nevertheless, I have done my best to organize my ideas and beliefs about Lee Je-ma's works that I've acquired through my personal experiences in the classrooms and clinical settings. I hope this book will inspire people to learn more about Sasang Constitutional Medicine and help it become a more valued asset in Korean Medicine.

I would like to express gratitude to all who have helped me in writing this book: Professor Song Il-byung and Professor Koh Byung-hee who had guided me onto the right path in my studies, the people at Jimoondang who have been completely supportive in the publishing of this book, nutritional specialist Baek Mi-young who helped with the diet plans, Lee ju-il and Choi Eun-kyeong who helped with the translations, and Miss Wang who did the revisions. And last but not the least, I would like to dedicate this book to my cherished mother, my wonderful parents-in-law, and my loving family Jun-ho, Song and my wife.

2012, March
At the research office
Lee Eui-ju

Contents

Chapter 1

Introduction to Sasang Constitutional Medicine

1. The Western countries in the late 18th century: the revolution in medicine

Social turbulences often act as catalysts in igniting scientific breakthroughs. The science of medicine is not an exception. Western medicine had advanced tremendously in the late 18th century throughout the early 19th century—a time of revolution and war. The modernization of medicine began in France, forging the modern day medicine that relies heavily on science and technology.

In conventional medicine, pathological states were not analyzed with biopsies or chemical analysis, as done today, but rather were classified according to the symptoms that could be grouped together to point to a specific malady. And the pathological conditions were believed to be the results of an internal imbalance and not of external pathogens. This ideology had its roots in the prevalent philosophy of that time as did those of other traditional medical cultures around the world.

But political unrest in the late 18th century triggered a change of event that altered the underlying ideology and methodology of medicine. World War I and World War II that followed the revolutionary period also brought cataclysmic changes. The wretched conditions in the wars showed that most of the preexisting medical knowledge was ineffective. The physicians who held higher social ranks in medical circles of that time were of no use in the war. Epidemic outbursts and traumatic injuries made doctors throw away the holistic and subtle medicine of the past and opt for the solid and concrete surgical techniques. Consequently, the boundaries between internal medicine and surgical methods began to thin out, leading to a shift of paradigm in medicine. For example, they began to search for specific pathogens causing inflammation, ulcers, and tumors, denying the conventional theories of bodily fluids and humors.

European countries like France was going through industrial revolution and urbanization that resulted in social turmoils and created public hygiene problems. Efforts to resolve this produced advancements in sanitary science, another accomplishment that marks 18th century as an important period in medical history.

2. The Joseon dynasty in the 19th century: A time of famine, epidemics, and riots

The 19th century of the Joseon dynasty was plagued by famines and epidemics. 70% of the total population did not have enough land to support themselves. During the 200 years between the reign of King Hyeon-jong and King Cheol-jong, severe famine struck Korea 52 times, according to official records. The famines were most severe in the late 18th century through the early 19th century, attested by the books of that time describing famine strikes. The famines were nationwide, and the famine in the countrysides in particular were severe enough to call for governmental countermeasures every three to four years, 3.6 years on average. There are morbid records that show how severe the famines were: parents leaving children they could not afford to feed, parents tying up their children to trees to stop them from following, and even parents driven to eat their own children.

The cholera outbreak was the worst blast of epidemic in the late Joseon dynasty. The outburst of cholera in Japan and China was concurrent with that in Korea, and the disease had apparently spread from the north to south.

Between the reign of King Hyeon-jong and King Cheol-jong, over 100,000 people died of epidemics, and smaller outbreaks, numbering up to 60, continued incessantly. According to a record, hundreds of thousands of people died of cholera in 1821. Historians agree that the words "hundreds and thousands" are actual statistical numbers and not metaphorical, for there are records which describe a whole village being wiped out in an outbreak.

The severe famines, epidemics, and riots brought extreme destitution for the people of the early 19th century, a time befittingly called "The Period of the Riots."

3. Confucianist medical doctors

Confucianist medical doctors were members of the nobility who chose to study medicine, who suddenly rose in number in the 18th century. Before this, medicine was considered a lowly art, an opinion that changed only in the late Joseon Dynasty.

Some historians claim that the nobility were forced to study medicine against will because of their degradation in power and wealth. But others believe that the social conditions of that time, with thousands of people suffering from famines and epidemics, could have moved them to realize the value of the medical profession.

In the same context, Jeong Yak-yong (1762-1836), the renowned scholar and author of 『Magwa hoetong』, introduced jennerization into Korea. Learned men of that time like Jeong Yak-yong who could see the devastations around them sought for ways to promote public health. Lee Je-ma, who had spent his entire life searching for ways to improve public well-being, probably took interest in medicine for similar reasons.

4. Lee Je-ma

Lee Je-ma

Lee Je-ma (1837-1900) lived in the late Joseon dynasty. He was born in the late 19th century and died at the break of the 20th century. He lived through the "Period of the Riots" of the late 19th century and was one of the people who experienced the up-heavals of society.

His given name was Dong-mu, and his appellation was Seop-wun or Seop-jin. His grandfather, Chung-won, and his father, Ban-oh, were descendents of the royal family, and Lee Je-ma himself was the 19th descendent of Ahn-won Dae-gun, son of King Mok-jo. Lee Je-ma was born on March 19, 1837 (on Korean lunar calendar), between Ban-oh and his mistress.

There is an amusing anecdote about his birth. One day, Ban-oh visited a tavern and became heavily drunk. The owner of this tavern had a plain looking daughter with such a difficult temperament that her parents could not find anyone to marry her. When Ban-oh stayed for the night, the owner of the tavern furtively sent his daughter into his room. Ten months later, Ban-oh's father had a dream of someone bringing in a swift steed from the Jeju island. Days later, a woman came into his house with a baby in her arms, and the baby was named Je-ma (meaning "horse of the Jeju Island").

Although he was an illegitimate child, he was treated no differently from his siblings. He received excellent education and left influential works in philosophy and medicine. Above all, he discovered the Sasang Constitutional Medicine.

1) Governmental Post

Lee Je-ma initially began his career in the military. He first became an official in 1875, at 39, and was promoted several times to different positions during the next eleven years.

In 1896, large scale riots led by Choe Mun-hwan stirred up nationwide disorder, for which Lee Je-ma was appointed head of the government force. He was again promoted in the following year but retired in 1898.

In his prime, he mostly stayed near Hannamsan or Hanbuksan, Gyeonggido. His mother became ill in 1895, precipitating his return to his birthplace. He opened a clinical practice in his later years in Mansegyo, Hamheung.

2) Scholarly works

There are descriptions of Lee Je-ma being strong-willed and decisive in temperament, and he won great love from his grandfather. Known to be a clever young man, he educated under Jik-Jang, his uncle. He discovered his bent for the martial arts and decided on a military career. His given name Dongmu means "warrior of the East."

He left home at the age of thirteen, travelling to far-away places to expand his experiences. In his youth, he travelled to Manchuria and the Soviet Union. He came across certain rare books in possession of a wealthy man named Hong, and it was there that he came in contact with the Western culture. Records tell us that, around 30, he chanced on 『Myeongseonrok』, a book written by Han Seok-ji (1709-1803), and was thoroughly riveted.

Lee Je-ma's philosophical viewpoints are revealed in his book 『Gyeokchigo』.[1]

The book consists of three chapters: 「Yu-ryak」, 「Ban-seong jam」, and 「Dok-haeng pyeon」. 「Yu-rak」, written between 1880 and 1889, attempts to reinterpret the classical confucianism in terms of Sasang Constitutional Typology. 「Dok-haeng pyeon」, written in 1882, proposes deeper knowledge and understanding of humanity as a solution for discordant relationships. 「Ban-seong jam」, written in 1893, expands on the same subject.

Lee Je-ma's philosophical ideas essentially boil down to "understanding the human being and rectifying the self." This forms the groundwork for his proposition that the mind needs to be healed in order to heal the body, expressed succinctly in the phrase "heal the mind to heal the body."

「Guangjeseol」 and 「Jejungsinpyeon」, writings on the preservation of health, and 『Dongmuyugo』, a recently rediscovered book, are all monumental works.

3) Medical studies

Lee Je-ma himself had suffered from severe illnesses and was diagnosed with "Yeolgyeok" and "Heyeok," pathological conditions in Korean Medicine that involve throwing up and coughing. This is probably one of the reasons he initially took interest in medicine. Apparently he theorized the constitutional typology while reading the books previously mentioned. Other than that,

---

1 『Gyeokchigo』
This book best illustrates his philosophical groundwork in the Sasang Constitutional Medicine. Here, he applied the Sasang philosophy, which deals with the four phasic elements of natural phenomena, to both the natural environment and the human society. This and 『Dongeui Soose Bowon』 are Lee Je-ma's most representative books.

it is hard to find evidence of teachers who could have taught him on the art of medicine.

There are records in 『Dongeui Soose Bowon』[2] that recount of a successful treatment of Soyangin with Yukmi-tang and Baekho-tang. Though he was inexperienced at that time, the backbones of his theories were apparently under construction.

He started writing 『Dongeui Soose Bowon』 at the age of 57 in 1893 and completed it a year later, at 58. He worked on a revised edition in 1899-1900, accruing more knowledge and experience. This book is considered the archetypical work in Sasang Constitutional Medicine because it lays the foundations of all the core concepts, explaining the physiology, pathology, pathophysiology, and treatment and prevention of disease for each constitutional type.

Recent studies have shown that Lee Je-ma had written yet another book, 『Dongeui Soose Bowon Sasang Chobon-gwon』.[3] Further research on the book is currently under way.

---

2 『Dongeui Soose Bowon (Longevity and Life Preservation in Eastern Medicine)』
Pertains to the printed script containing 4 books in 2 volumes. Initially, only a part of the whole work was printed in 1894, and extended editions were printed afterwards. When Lee Je-ma died in 1901, two of his students, Kim Yeong-gwan and Han Mok-yun, gathered the remaining material and published the final version.
This book contains the complete ideas on Sasang Constitution. It consists of seven chapters: 「Sungmyeong」, 「Sadan」, 「Hwakchung」, 「Jangbu」, 「Euiwon」, 「Gwangje」, and 「Sasangin byeongjeung」. Every chapter carries detailed explanations of the clinical approach and the medicational system found in Sasang Constitutional Medicine. The book introduces a system of constitutional typology which categorizes people into four types of constitution (Taeyangin, Soyangin, Taeeumin, Soeumin). Its medical system focuses on the type of constitution, rather than on the manifested symptoms. In contrast to the classical Korean Medicine, which has its roots in the Yin Yang and Five Elements theory, Sasang Constitutional Medicine emphasizes the importance of the constitutional type. Medical historians view this shift of paradigm as a milestone in the history of medicine.

3 『Dongeui Soose Bowon sasang chobon-gwon』
The author of this book remains yet to be identified, but specialists who have examined the text and the written style believe this script to be one of Lee Je-ma's earlier works, written before 『Dongeui Soose Bowon』. It contains writings on his theoretical groundwork, preventive medicine, and the earlier Sasang Constitutional medicaments.

5. Constitutional Medicine

Constitutional Medicine is a science that studies the physiology and pathology of the four constitutional types. The concept that constitution can be "typed" is the underlying paradigm evident in 『Dongeui Soose Bowon』.

Lee Je-ma's medical system is termed "Constitutional Medicine" because it is rooted in the concept of constitutional typology, a unique methodology of typing people according to their constitutional makeup. His ideas were shaped in the 19th century, a time of much sociopolitical unrest. Perhaps because of the great social demand for change, the underlying philosophy observed in Constitutional Medicine is also unconventional in view of the traditional confucianist beliefs.

『Dongeui Soose Bowon』 was written after 『Gyeokchigo』 was completed. In the book, he continues to expound on his ground-

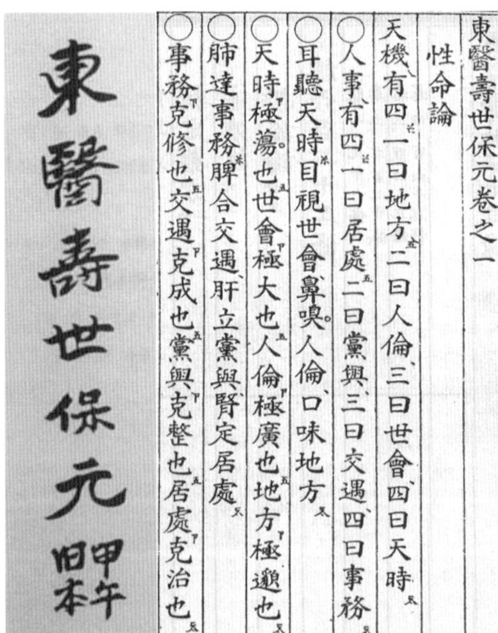
東醫壽世保元卷之一
性命論
天機有四一曰地方二曰人倫三曰世會四曰天時
○人事有四一曰居處二曰黨與三曰交遇四曰事務
○耳聽天時目視世會鼻嗅人倫口味地方
○天時極蕩也世會極大也人倫極廣也地方極邈也
○肺達事務脾合交遇肝立黨與腎定居處
○事務克修也交遇克成也黨與克整也居處克治也

『Dongeui Soose Bowon』

breaking ideas. Constitutional Medicine does not focus on the Yin Yang theory or the Meridian System theory, but rather concentrates on the type of constitution, explaining the physiology, pathology, and treatment and prevention of disease peculiar to each type of constitution.

Lee Je-ma reveals the new medical paradigms he discovered while studying conventional medicine. He also speaks about the medical figures and written works that had influenced him the most.

Information on the people and works that contributed to the shaping of his theories will help us understand Constitutional Medicine better. One of the people he had a very high opinion of was Zhang Zhongjing.

Zhang Zhongjing is the author of the classic script 『Shanghanlun (Treatise on Cold Damage Diseases)』. 『Neijing (Internal Classic)』,[4] the archetypical book from which all of traditional medicine stems, proposes imbalance within the body and disharmony with nature as the most important pathogenic factor; the book provides the most fundamental philosophical background for theories on treatment and prevention of illnesses. On the other hand, 『Shanghanlun』 deals with the symptoms of and treatment for specific disorders; this book instructs more concrete and detailed ways for treating patients.

Lee Je-ma's medicine is more closely related to the ideas presented in Shanghanlun than those of Internal Classic, suggesting

---

4 『Neijing (Internal Classic)』

Also called Naegyeong. It burrows a form of dialogue between Hwangje (a mythical ruler of China), the purported author, and Gi-baek (a retainer at the royal court) in narration, but historians believe that the real author wrote the script under a pseudonym. The work consists of 18 books, divided into two volumes of nine books each, and the two volumes are titled "Somun" and "Yeongchu," respectively. Somun elaborates on the natural laws in traditional medicine, such as the Law of Harmony Between Man and Nature and the Law of Yin and Yang, but does not contain the actual clinical procedures. Yeongchu contains material on acupuncture and moxibustion, but explains little on herbal medicament. The surviving edition of 『Neijing (Internal Classic)』 annotated by Wang-bing in the Dang era consists of twenty four books. There was also a 30 book edition edited by Yang-sang-seon in the Su era, which is no longer existent.

that he was more interested in the practical use of medicine. Most students of Constitutional Medicine follow this inclination, focusing more on practical clinical application than on theoretical studies.

Descriptions of the important personage and books in the history of practical medicine appear in various parts of this book. Also mentioned are some of the writers whose works are often quoted by Lee Je-ma. There is also a section introducing the esteemed physician Heo Joon and his groundbreaking work, 『Dongeui Bogam』.

6. The influences of 『Dongeui Bogam』

In Lee Je-ma's estimation, the completion of 『Dongeui Bogam』 marked the height of the renaissance in Korean Medicine. He held a high respect for its author, Heo Joon, matching that for Zhang Zhongjing.

Heo Joon

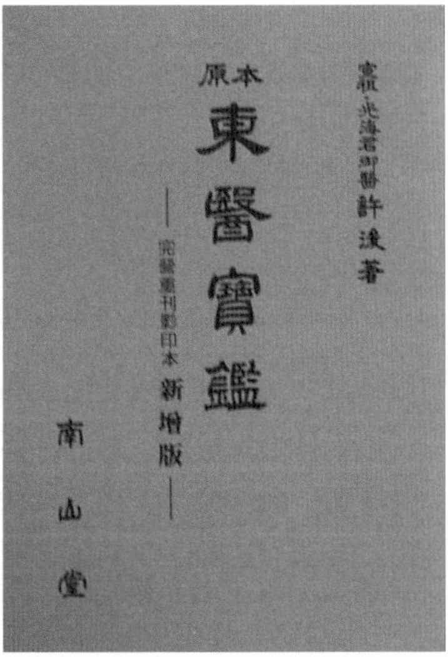

『Dongeui Bogam』

Lee Je-ma refers to 『Dongeui Bogam』 frequently in his writings, and the book clearly helped shape much of his theories.

He quotes passages from 『Dongeui Bogam』 in his descriptions of pathological conditions, criticizing the conventional medical views and proposing a new perspective.

In his comments on preventive health care in 『Dongeui Soose Bowon』, he again burrows ideas from 『Dongeui Bogam』, emphasizing Taoism and the important role of moral integrity for a healthy life.

7. The characteristics of Sasang Constitutional Medicine[5]

The differences between the classic Korean Medicine and the Sasang Constitutional Medicine are as follows:

First, Sasang Constitutional Medicine sees each person as an individual with distinct constitutional traits.

Up to then, traditional medicine assumed that the physical body operates in the same way for everyone. But in Constitutional Medicine, the four constitutional types react differently to different situations, both physically and psychologically. This gives rise to distinct patterns of pathology, and the respective medications, found exclusively in a specific constitutional type. The medicational reaction is also different for each constitutional

---

5 The law of Yin and Yang states that the universe, both the microcosm and macrocosm, is always composed of two contrasting sides, the Yin and the Yang—the dark and light, the cold and hot, the heavy and light. When this binary aspect is once again bifurcated, it results in the Sasang (四象), or the Four Reflections—the four aspects of a single phenomenon. The Yang (陽) bifurcates into Taeyang (太陽) and Soyang (少陽), and the Yin (陰), or Eum, into Taeeum (太陰) and Soeum (少陰). Lee Je-ma's constitutional medicine is termed "Sasang Constitutional Medicine" because it is a system of constitutional typology that divides people into four types according to the constitutional makeup, namely Taeyangin (太陽人), Soyangin (少陽人), Taeeumin (太陰人), and Soeumin (少陰人).

type. He thusly emphasized the idiosyncratic qualities of the pathological patterns found in each type of constitution.

Second, Sasang Constitutional Medicine is an idiosyncratic medicine.

Constitutional Medicine draws attention to the patient's idiosyncratic reaction to medication. Usually, patients do not recover uniformly under the same treatment. For one thing, the time required for adequate treatment differs among people. The Yang type (Taeyangin and Soyangin) respond quickly to medication, whereas the Yin type (Taeeumin and Soeumin) require a longer period of time. Also, some constitutional types are more prone to certain patterns of pathological progression. The Yang type tend to show a rapid progress of disease, while the Yin type display a slower one. Therefore, it is necessary to take these factors into account and watch out for the idiosyncratic patterns.

Western therapy focuses on the "pathogen" compartment of a disease. For example, if a patient is diagnosed with pneumonia, he is treated with antibiotics that can exterminate the pathogenic agent. On the other hand, Eastern therapy focuses on the "host" compartment of disease. We do not exterminate the causative agent directly, but rather encourage the host to recover and overcome the pathogenic factor on his own. In other words, the pathogenic agent is indirectly incapacitated as the symptoms are being treated. Furthermore, Sasang Constitutional Medicine provides specialized therapy for each type of constitution, which is why it can be called an "idiosyncratic" medicine.

Third, Constitutional Medicine asserts that the body is healed through the healing of the mind.

According to traditional medicine, most illnesses fall into one of the two categories: disorders from external causes (eg. contagion, inclement weather, physical trauma) and disorders from internal causes (eg. unbalanced diet, unhealthy sex life). Sometimes, they include another pathogenic source (having much less significance than the first two): unstable emotions and unrequited desires. By contrast, Lee Je-ma perceived and emphasized the innate differences between the constitutional types in the organ structural schemes of the body, or the structure of organ systems specific for each constitutional type. Even after birth, a person's physical conditions are constantly being affected by the mind. Therefore, controlling the mind is important in the prevention and treatment of disease. Each constitutional type has different desires and tendencies, which is why the mind should be cultivated differently for different constitutional types.

8. Constitutional Medicine of the past and present

In the 19th century, Lee Je-ma, both a philosopher and a physician, proposed the Sasang Constitutional Typology and its related theories. The concepts he presented were based on the mainstream Confucianist doctrine in Joseon, and he also strived, as a medical practitioner, to provide as many people as possible with the benefits of medicine.

The foremost objective of his philosophy and medicine was promoting public well-being, in both the mind and the body, and protecting the general populace from deadly epidemics. In 1894, his efforts produced the book 『Dongeui Bogam』, which was revised several times afterwards as he gained more experience, until his death in 1900.

After his demise, his followers published a revised edition of the 『Dongeui Bogam』, which incorporated the first version (written in 1894)

『Dongeui Soose Bowon』 published in 1901

and the most recent one (completed in 1900). Han Du-jeong, one of his students, printed the 7th edition in 1941, adding more herbal medications and presenting the material in a simple text that even a layman could comprehend. In 1945, Choi Seung-dal and Lee Hyeon-jae established the Institute for Promotion of Sasang Constitutional Medicine. Hong Soon-yong and his followers launched the Society of Sasang Constitutional Medicine in 1970, setting off a full-scale research on Constitutional Medicine. Later in 1945, this organization officially became a subdivision of the Korean Oriental Medical Society. Regular meetings and conventions were held beginning in 1979. In 1989, the first issue of *Journal of Sasang Constitutional Medicine* was published, a collection of papers on theories, acupuncture, medication, preventive care, and scientific experiments on Sasang Constitutional

Lee Hyeon-jae (2nd) and Park Seok-eon (6th) in front of Myeong Church in 1950

The first conference of the society of Sasang Constitutional Medicine (May 30, 1970 AKOM news)

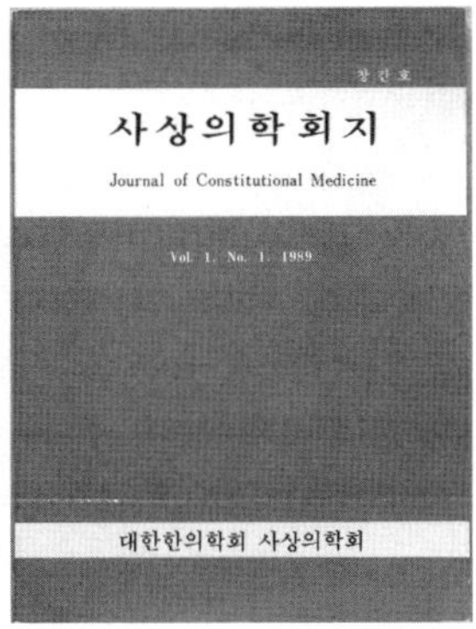

The first issue of Journal of Sasang Constitutional Medicine (1989)

Medicine, now printing three issues every year. The Society of Sasang Constitutional Medicine has become the center of research on Sasang Constitutional Medicine for students, researchers, and practitioners alike.

(For more information, visit The Society of Sasang Constitutional Medicine website at http://www.esasang.com)

9. Medicine that heals the mind

The Western medicine has advanced to the point where a person's life can be predicted and analyzed through genetic analysis. Moreover, Western medicine displays certain advantages over Eastern medicine; for example, Western medicine had discovered cures for cholera and smallpox, which had killed so many in Korea in the Joseon dynasty.

Western medicine split its bond with natural philosophy in the late 18th century, and the breach perpetuates to this day. This was probably an inevitable event in the course of history, for the cultural current was flowing away from the mystical and philosophical to the scientific and mathematical.

But also, because the human being is composed of the mind as well as the body, medicine needs to embrace the mental part of a man. This requires a perception of the mind and body as intricately interrelated elements of a homogeneous entity, a concept wholly different from the mind-body dualism paradigm prevalent in modern day Western medicine.

Morals and ethical standards are constructed at the same time as a culture is being developed. Once the cultural code is securely instilled, the members of the society adhere to these standards without question, a tendency that explains the impact of culture on the mind and body of a man.

And so, in order to encompass the psychological part of a man, the Western culture and medicine need to attain a better understanding of psychology, philosophy, and morality. In contrast, the Eastern medicine, especially the Sasang Constitutional Medicine, had always retained its connections with the human mind.

The virtues of this quality is not fully realized in today's society, possibly due to a lack of understanding on Eastern medicine and its consequent alienation from the predominant (Western) medicine. As proof, the Korean people mostly associate Korean Medicine with herblore and natural remedies, at best visualizing it as a form of preventive medicine. And so, further studies and efforts are required in order to promote Constitutional Medicine and present the public with a more comprehensive medical care.

Chapter 2

Determining the constitutional type

1. Understanding constitution

"Constitution" is a frequently misused term. Some people use it as in "a peculiar constitution" or "an allergic constitution." It is sometimes used to refer to the "Sasang Constitution," at other times used to describe a "weak constitution." It's a word that people have a hard time understanding, and yet at the same time are familiar with.

In this chapter, we'll go over the concept of "constitution" and identify the different versions of constitutional typology.

Constitution can be defined as "the aggregate of physical form and functional characteristics that is shaped by the hereditary determinants and environmental factors."

The definition of constitution can differ slightly from person to person depending on his main field of study. Constitutional typology is a part of the medical tradition that had been received with much attention and research in countries both Eastern and

Western.

In the Western world, Hippokrates[1] proclaimed 'the theory of the four humors.'[2] Later on, Galenus[3] added a new concept to

1 Hippokrates (B.C. 460? - B.C. 377?)
Born on the Greek island of Kos, he received medical training from his father. Later on, he acquired more medical knowledge through encounters with other philosophers and doctors during his travels in Asia and other parts of Greece. He returned home and opened his practice, working on his books at the same time. He then traveled again, expanding his thoughts and widening his vision. The Corpus Hippocraticum is an anthology written by several authors, including Hippokrates and his students (in which Hippokrates's voice is easily discernible).
His views on the physiology and pathology of the human body are rooted in his 'theory of four humors.' He observed the symptoms of his patients during therapy and concluded that pyrexia (feverish states) indicates recovery from a disorder. He referred to physis, or Nature, as that which heals, the vital life force inherent in the human body that allows us to maintain homeostasis and a balanced internal environment. In his therapeutic methodology, he contested that the physis must be supported without impeding on its natural mechanisms.
He also studied symptomatology and prognosis, and his portrayal of near-death patients are acceptable even by modern standards. He emphasized the code of professional ethics and has won the honorable title of "Father of Medicine."

2 Theory of the four humors
This theory, advanced by Hippokrates, are the fundamentals of his views on human physiology and pathology. Rooted in the ideas of Essentialism, it postulates that the human body contains certain mixtures of the four elements (earth, air, fire, and water) called humors (blood, phlegm, black bile, and yellow bile). Health was defined as the balance of the four humors, or 'eukrasie,' and disease was defined as the imbalance, or 'dyskrasie.'

3 Galenus, Claudius (131-201)
This physician was born in Pergamum, Asia Minor. His father was an architect who was well-versed in mathematics, natural sciences, and philosophy, which he taught to his son. Galenus's medical training began at 17, pursuing further studies in Izmir, Corinth, and Alexandria. At 28, he returned home to open a practice. He became a renamed figure by the age of 34, befriending some prominent scholars and high officials. But he also antagonized many others, which forced his return to his homeland in 166.
Later, he followed an army led by the Roman emperor Marcus Aurelius, and on his return to Rome, he was appointed court physician to the emperor's heir, Commodus. His manuscripts from that time period explores beyond the boundaries of medicine to philosophy, linguistics, and mathematics, and his works on anatomy and physiology won much acclaim.
He drew conclusions on the structure of the human body through physiological experiments and studies on zootomy, forming the scientific footholds of his medicine. His compositions, together with those of Hippokrates, made critical impact on the shaping of Western medicine. Whereas Hippokrates emphasized empirical medicine, Galenus's medical paradigm was wholly rooted in teleology, and he derived his

the theory, that of the four temperaments[4] affecting the humors. Kretschmer[5] divided the general population into three psycho-somatic types,[6] while Jung[7] proposed the theory of psychological

---

theories from natural phenomena that he believed conform to a ultimate purpose.

4 Theory of the four temperaments

According to Galen, there were four temperaments associated with the humors (sanguine, phlegmatic, choleric, and melancholic) producing different moods in people. Sanguine temperament is warm hearted, sociable, passionate, bright, easily excited; phlegmatic temperament is cold, calm, stubborn, strong willed; choleric temperament is impatient, emotional, bold, objective; melancholic temperament is strong willed, persistent, easily depressed, subjective, conservative.

5 Kretschmer, Ernst (1888. 10. 8 - 1964. 2. 8)

German psychologist, born in Wűstenrot. He became the director of the psychiatric clinic at Marburg university and the University of Tűbingen in 1926 and 1946, respectively. He wrote *Der sensitive Beziehungswahn* (The sensitive relative delusion) in 1918 and *Kőrperbau und Charakter* (Physique and character) in 1921, and become a prominent European figure in psychiatry and psychology. He introduced into the German psychiatry a dynamic paradigm that emphasized the role of environmental factors, giving rise to the idea that temperamental reactions are reflections of the somatotype.

6 Three physical types of Kretschmer

Kretschmer attempted to correlate body build and physical constitution with character and mental illness. Through research on personality traits found among schizophrenia and manic-depression patients, he established a binary division: the "schizoid" and the "cycloid." Kretschmer studied physical and psychological typology in his medical experience, especially the morphological-physiological-psychological uniformity of the individual. He identified the physical types—the pyknic (rotund, or endomorph), the athletic (muscular, or mesomorph), and the asthenic (tall and thin, or ectomorph)—claiming that different psychiatric disorders are associated with each type (endomorph—manic-depressive; mesomorph—epileptic; ectomorph—schizophrenic).

7 Jung, Carl Gustav (1875. 7. 26 - 1961. 6. 6)

Swiss psychologist, born in the small village of Kessewil. After graduating University of Basel, he studied psychiatry under Eugene Bleuler, a professor at University of Zurich, at the Burghoeltzli Mental Hospital in 1900. In 1904, he was the first to recognize the efficacy of analytic psychology, inventing the word association test, and he validated Freud's theory on the suppression of unconscious, coining the term "complex." In 1906, he proceeded to prove the efficiency of the psychoanalytic method in understanding schizophrenia. These achievements won Freud's confidence, who eventually came to see Jung as his heir apparent.

In April of 1908, he was elected editor of the Jahrbuch, a journal printed by the International Association of Psychoanalysis, of which he was appointed the first president. However, his theoretical ideas on "libido" diverged acutely from that of Freud, leading to his resignation in 1914. He founded his own movement (analytic psychology) in later years, but the insinuations evident in his works about spirituality and the complexity of the material undermined his potential influence on psychology in general. Nevertheless, his works on extroversion and introversion won much acclaim.

archetypes in mankind, namely the "Jungian archetypes." In modern immunology, four different types of hypersensitivity are identified in clinical settings (types Ⅰ, Ⅱ, Ⅲ, and Ⅳ).

In contrast, in the Eastern world, the concept of constitutional typology starts off in Neijing (Internal Classic), the archetypal book in Eastern medicine, which mentions "the five type of people."[8] The Sasang Medicine, a form of constitutional typology[9] introduced in Lee Je-ma's 『Dongeui Soose Bowon』, is currently in use in Traditional Korean Medical practice.

Today, the typological system remaining in the Western medical tradition, such as those used in psychiatry or immunology, focuses more on the observable patterns than on the internal constitutional makeup. By contrast, the system of typing applied in the Eastern medical practice conforms to the constitutional typology found in Sasang Constitutional Medicine, which incorporates both the physical and psychological characteristics.

The following sections will provide a cursory look into the constitutional types identified in Korean Medicine. Lee Je-ma classified people into one of the four types of constitution (Taeyangin, Soyangin, Taeeumin and Soeumin) and explained the different physiology, pathology, and beneficial preventive health care respective of each type's physical and psychological traits. The branch of Korean Medicine that studies the constitutional typology is called the "Sasang Constitutional Medicine."[10]

---

8 Theory of the five types of people

This is a system of constitutional typology introduced in Naegyeong that is rooted in the Yin Yang Theory and the Five Phasal Elements Theory. The five types, Taeyangin, Soyangin, Taeeumin, Soeumin, and Eumyanghwapyeongjiin (the equilibrated type), are classified according to the temperamental disposition. This typology system does not take into account the physical build or the internal organ makeup, conferring limitations in the treatment and prevention of disease.

9 Constitutional Typology

The two most important elements in the Sasang typology system are Seong and Jeong. Seong is the innermost nature of the minds; Jeong is the outward emotional disposition.

10 Sasang Constitutional Medicine

Also termed "Constitutional Medicine," this system of constitutional typology identifies four constitutional types—Taeyangin, Taeeumin, Soyangin, and Soeumin—each

I have strived to accurately recount the physical forms, psychological patterns, beneficial regimen,[11] and proper health care for the four constitutional types that Lee Je-ma described in his unique Sasang Constitutional Typology system.

2. Which constitutional type am I?

In my office at the clinic, I often encounter patients who ask me to "change their constitution."

According to Constitutional Medicine, once the constitutional type is decided, it can't be changed. Each constitutional type has its particular physiological and pathological characteristics that work from completely different dimensional planes. This is called "the principle of constitutional invariability."

What is changeable is the imbalance within the constitutional type. The regulation of this imbalance, done by replenishing the insufficiency and removing the excessiveness, is called "improving the constitution." The constitutional type and the constitutional health need to be differentiated in order to avoid confusion while studying constitutional typology.

Personally, I routinely examine the three following criteria to determine the constitutional type of a person.

The first criterion is the outside (body), or the appearance,

---

associated with a particular pathology and therapeutic approaches.

11 Beneficial regimen

The Eastern studies on beneficial regimen precede modern nutrition and dietetics by 2000 years. Somun, the first of the two volumes of 『Neijing (Internal Classic)』, contains a deluge of information on the basic medical concepts (physiology, pathogenesis, theoretical backgrounds) and beneficial regimen. Despite the popularity of Confucianism among the learned men of that time, shamanistic folk remedies were more common among the commoners, which set longevity and eternal youth as the paragon of happiness. These beliefs bonded with primitive Taoism and coalesced into native folklore, appealing strongly to the Chinese people. They were reinterpreted through the teachings of Lao-tzu and Chuang-tzu and were subsequently merged into religious Taoism.

specifically the facial features and somatotype. For example, at the fruit shop, we can discern the apples, oranges, or watermelons by their shape and color, because we already have an idea of how they look. In the same way, the constitutional types in Sasang Constitutional Medicine each have particular physical qualities that are manifested externally, which can allow easy self-diagnosis of our constitutional type.

The second criterion is the inside (mind), or the basic mind-set, specifically the temperament, caliber, natural disposition, and avaricious desire. For example, one can identify apples, oranges, or watermelons by their taste and smell without actually seeing them. In the same way, the constitutional types each have unique psychological traits that set them apart, an important concept in Sasang Constitutional Typology.

The third criterion is the symptomatic pattern. This includes the physiological patterns, noncritical pathological symptoms, and critical pathological symptoms. Because each constitutional type has different physiological conditions, the pathological patterns also manifest differently. This third criterion demands medical expertise, rendering it inapproachable for the layman.

The three criteria mentioned above the outside (body), the inside (mind), and the symptomatic pattern are the key criteria in constitutional typing. However, in Eastern medicine, it is also understood that the part can have meaning only within the context of the entire frame. Since each criterion taken separately is inadequate for a comprehensive assessment, all three criteria must be analyzed together to produce an integrative picture.

Even today, the basic standards in constitutional typing are the external appearance, internal mind-set, and symptomatic pattern. The routine procedure in constitutional diagnosis is to first observe the facial features and posture, then to check the patient's lifestyle and personality traits using the questionnaire form

Sasangin Diagnosis Method

SF_SDQ (Short From_Sasangin Diagnosis Questionnaire)

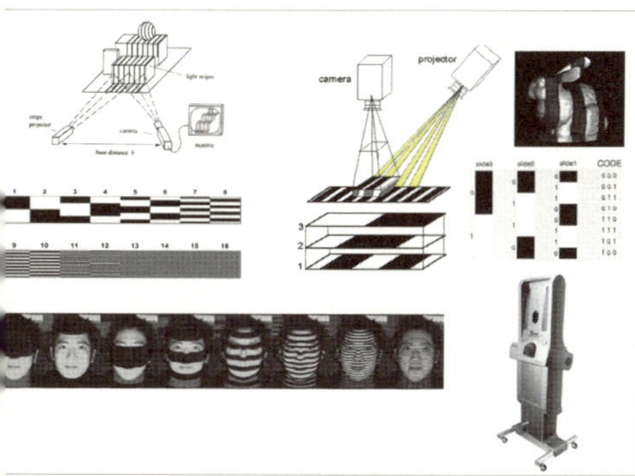

Sasangin Face Analysis

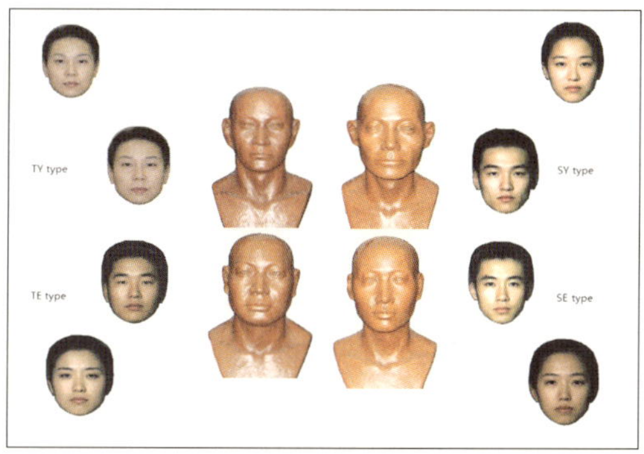

Sasangin Standard Face

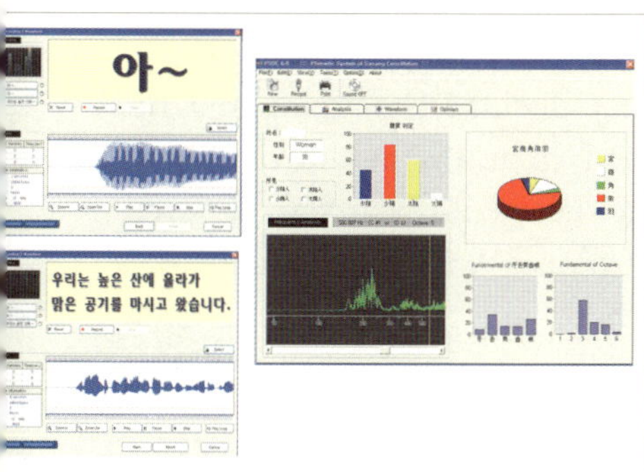

Sasangin Voice Analysis

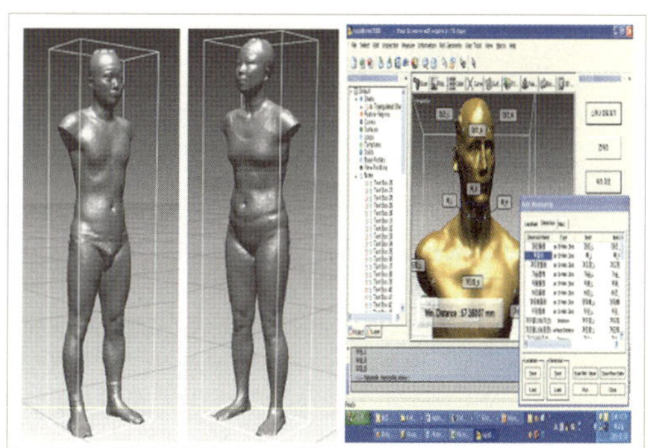

Sasangin Body Frame Analysis

(QSCC II),[12] and then to question the past medical record and family history. With the constitutional type thus decided, the patient's reactions to herbal medication and acupuncture therapy are assessed to confirm the final diagnosis. Unfortunately, this method of constitutional diagnosis lacks sufficient objectivity, which is why some colleges and other organizations have collaborated for the past ten odd years to develop more objective diagnostic tools. At first, there were attempts to standardize constitutional diagnosis by morphological characteristics (eg. cephalic morphology, breast size, shape of the ilium, etc.) There were also researches involving heat metabolism, hematogenesis, and immunological theories. In one project, digital instruments were used to standardize the facial contour. Researchers applied the rapidly advancing genetic engineering technologies in the most recent and highly anticipated project that uses DNA chips for constitutional typing.

3. What is the Taeyangin (Taeyang type) like?

A 15 year old maiden visited Lee Je-ma one day on account of frequent epistaxis (nosebleed). He interviewed and touched her, but unable to decide on her constitutional type, he took her on a walk. They trekked 5ri (1ri approximates 3.92727km), throughout which he observed her gait and behavior, before he determined her to be a Taeyangin. That decided, he instructed her to ingest a boiled decoction of reed roots on her next episode of nosebleed. She was also told that she would bear few children.

Lee Je-ma himself was often perplexed with the complexities of constitutional diagnosis, and there are traces of him over-

---

12 QSCC II (Questionnaire of Sasang Constitution Classification II)
This diagnostic tool was invented in 1996 in the department of Sasang Constitutional Medicine, College of Korean Medicine, Kyung Hee University. It is a computerized version, the validity of which has already been verified through several tests.

coming the obstacles of constitutional typing through painstaking observation.

Taeyangin constitute only 0.3% of the entire population, and this scarcity increases the difficulties of discern ing their constitutional type. The most visible Taeyangin attribute is that they have an excessively developed upper burn-center—head and upper chest—and a poorly developed the mid-lower burn-center—flank and lower back.

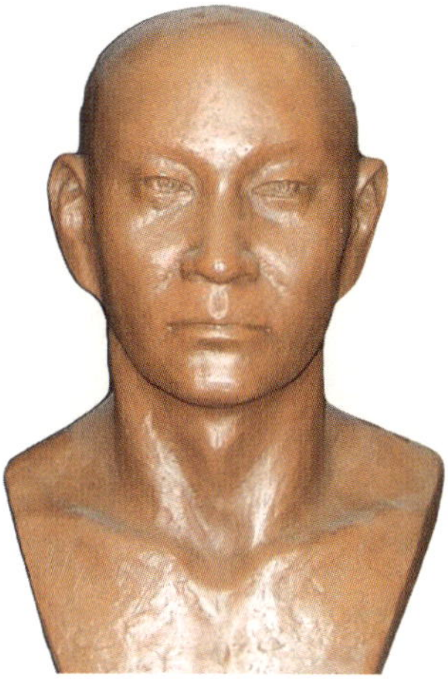
Taeyangin

1) External attributes (body)

The facial features are regular, with a wide forehead, high-bridged nose, big ears, and characteristically shiny eyes. They possess a large-headed and thick-necked body frame, with broad shoulders and back, short and narrow waist. Their small hips and skinny legs confer them an unstable standing posture, and they in fact do not enjoy standing or walking around at length. Taeyangin women are healthy and sturdy in appearance, but their uterine functions, which in Korean Medicine are related to the mid-lower burn-center—the lumbar parts—are weak, a condition that lead to possible infertility.

2) Internal attributes (mind)

The intrinsic qualities are reflected in many aspects: temperament and calibre (capability, talent, strength and weakness), natural disposition (the nature state of mind), personality (Seong and Jeong), overweening desire (greed rooted in an untempered mind), etc.

(1) Temperament and calibre

Taeyangin are communicative and decisive, and their sincerity and integrity bestows them the ability to discern easily the good and evil in a man. They are gifted with the social grace that allows them to interact fluently in relationships. They reach out readily and connect easily, and are rarely diffident or reserved. Nevertheless, they are strongly averse to forming close knitted cliques, and are particularly intolerant to deception and connivery, becoming uncontrollably enraged on such occasions.

(2) Natural disposition

Taeyangin are often impatient and short-tempered in nature. This comes from being "unable to attain everytime what was sought after," which in turn causes frustration and vexation. According to the Sasang Constitution theory, this impatience needs to be curbed in order to allow for a flawless functioning of the hepatic system. When life runs smooth, the Taeyangin effectively controls his temper and retains placidity; when life turns rough, his Hang-shim gains ascendancy and debilitates his physical conditions.

(3) Personality

The outbound facet, which arises from the Seong or the intrinsic selfless psyche, of the Taeyangin personality always seeks to move forward and refuses to retreat. They thence tend to emphasize logical causality and are determined to perfect their lives. A step further from this stage will turn them into obsessive perfectionists.

The inbound facet, which arises from the Jeong or the extrinsic self-centered psyche, of the Taeyangin personality always leans toward masculinity away from femininity. They deny any methodology other than their own, always having to win the first place. When this runs out of control, they enter the following state of overweening desire.

(4) Overweening desire

The overweening desire of Taeyangin are manifest in the form of dissoluteness. When the propensity for masculinity becomes excessive, self-indulgence and dissoluteness begins to materialize. The Taeyangin then sees only advance and no retreat, forcefulness only and no leniency. When this mental attitude becomes more articulate, it is expressed in the form of dissoluteness, and the Taeyangin turns exceedingly intractable, putting off his closest friends. The state of matters subsequently deteriorates into one filled with self-righteousness and bigotry.

As described above, Taeyangin can be determined leaders if all goes well, but they are equally susceptible to the lures of a dictatorship.

3) Symptomatic patterns

The following depicts the symptomatic patterns found in a physiological state (the healthy state free of pathological conditions), non-critical pathological state (the less sever forms of illness), and critical pathological state (the more severe forms of illness).

Unhampered urination is the cardinal indicator of the unaffiliated state in Taeyangin.

Non-critical pathological patterns are the less severe conditions. According to Sasang Constitutional theory, constipation that persists for 8-9 days can be treated lightly as long as there are no urinary problems. However, if saliva or foam gathers in the mouth, therapeutic procedures become necessary.

Critical pathological patterns are the more severe forms of illness. The overdeveloped state of upper burn-center induces Yeolgyeok and Banwi, whereas the underdeveloped state of the mid-lower burn-center induces Haeyeok. Yeolgyeok is a disease involving dysphagia or emesis due to gastroesophageal conditions. Banwi is a form of delayed emesis where the foodstuff taken in enters the stomach but is regurgitated after a certain period of time. Haeyeok is a motor disorder involving immobility caused by debilitation of the leg, but is unrelated to sensory disorders like paralysis or pain.

Taeyangin are mostly healthy, free of trivial sicknesses, one of those who never have to go to the hospital. But once he they become ill, they require prolonged treatment and convalescence.

4. What is the Soyangin (Soyang type) like?

According to an anecdote, an ailing maiden visited Lee Je-ma after numerous unavailing treatments by other physicians to cure her illness. Unable to determine her constitutional type in a cursory exam, he sent the rest of her family out of the room, and suddenly stripped her clothes off. To this the maiden resisted furiously, clawing and biting and screaming at the top of her voice. Upon this, Lee Je-ma smiled and explained that all her worries were over because her conduct had proved her as a Soyangin, after which he successfully cured the previously "incurable" disease.

Soyangin constitute 30% of the entire population and are the most easily distinguishable constitutional type.

1) External attributes (body)

The Soyangin have facial features that give a rather sharp impression, with uplifted eyes and ears, thin lips, and narrow chin. Conspicuous hammer-shaped and bulging skulls are commonly found in Soyangin, whose voices tend to be crisp and penetrating, their garrulity easily fostering misimpressions of flippancy. They are also careless speakers whose thoughtless remarks can put off their more sensitive friends. To give them their due, they are equally quick to forgive.

Their body build of the broad, well developed chest and the

skimpy lower limbs resembles an inverted triangle. The upper part of the Soyangin body is said to be relatively full of energy conducive to their characteristic agility and nimbleness, whereas the relatively lethargic lower part produces unstable sitting postures.

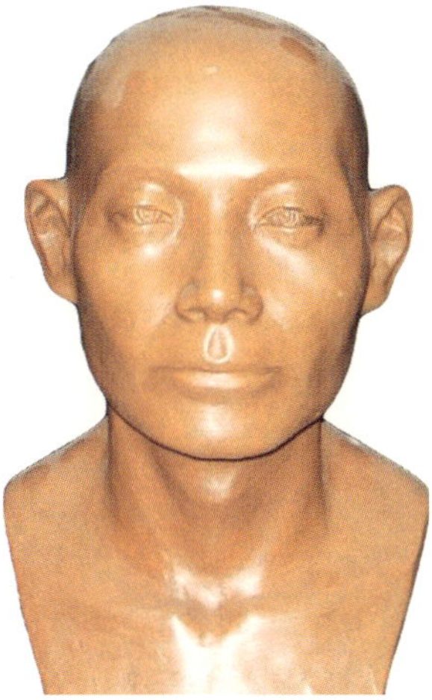

Soyangin

Even among Soyangin, there are some who are short in height and of delicate features that confers them a semblance of the Soeumin. Therefore, determination of the constitutional type must not depend entirely on appearances, but rather also on the inner nature of mind and the characteristic symptomatic patterns.

2) Internal attributes (mind)

(1) Temperament and calibre

Soyangin are forceful and strongwilled, and they easily perceive the knowledgeability or ignorance in a man.

They are men of worldly affairs who are gifted with efficacy and good task management skills, allowing them to recognize the structural sequences and interrelationships in matters, and thus work things through speedily. They are also receptive to new methodologies that help them attain greater efficacy.

(2) Natural disposition

A close look into the Soyangin mind reveals that it is in fact apprehensive, uneasy, anxious, and restless in nature, a state Lee Je-ma believed comes of being "unable to attain amply what was sought after." By nature, Soyangin tend to start out on things impulsively and lightly but are unable to stick through it, the ones who "spread out without pulling in." The inevitable backwash that results from this pattern of conduct instills apprehension and uneasiness on the mind. When the psychological pressure thus generated accumulates to a certain level, it produces states of dread and fear, sometimes developing into amnesic states, an expression of the non-critical pathological pattern.

(3) Personality

The outbound facet, which arises from the Seong or the intrinsic selfless psyche, of the Soyangin personality is always spreading things out and never pulling them in.

The inbound facet, which arises from the Jeong or the extrinsic self-centered psyche, of the Soyangin personality is always centered around the external matters and dismissive of the internal matters. The Soyangin thrives on any affairs other than those of his own family while his domestic life drifts apart unheeded. If this penchant is not kept within limits, the Soyangin then falls into the following state of overweening desire.

(4) Overweening desire

The overweening desire of a Soyangin are presented in the form of bigotry. The prejudiced mind is said to be conducive to the desire to take away from another. Subsequently, the Soyangin is no longer able to discriminate between the public and private affairs, nor between necessities and trivialities nor between the important and unimportant matters. Rather, they deal with matters impulsively and impudently, indulging in self-adornment, extravagance, and dissipation.

Soyangin are spirited, energetic, efficient, and extroverted, gifted with an unusually ingenuous, chivalrous, and honorable nature. Their apparently flippant demeanor does not undermine the selfless sympathy and kindliness evident in their conduct.

However, their shortcomings include a quickness of temper and a relative lack of interest in domestic matters compared to their voracious appetite for the "outside affairs." They are also lacking in tenacity, starting things off impulsively, but rarely pulling together in the end, becoming easily unfocused and quick to give up.

3) Symptomatic patterns

The following describes the symptomatic patterns in a physiological state, non-critical pathological state, and critical pathological state.

Unimpaired bowel movement is the prime evidence of a healthy Soyangin. The people who are prone to bouts of constipation in less healthier conditions are likely to be Soyangin. Unlike constipation in other constitutional types, Soyangin constipation progresses rapidly and thus demands extra attention.

Non-critical symptomatic patterns are the less severe forms of illness. When the uneasiness and apprehension, the natural disposition of a Soyangin, amplifies into more the insidious fear and dread, it can in turn lead to amnesic states.

Critical symptomatic patterns are the most advanced stages of illness. When the imbalances of the mind intensifies, the Soyangin becomes prone to spells of intense sorrow and grief, while the usual constipation can suddenly worsen, causing a fiery, congested sensation in the chest.

Other illnesses found commonly in Soyangin include urogenital disorders (especially of the prostate), peptic ulcer, neurological disorders (low back pain), and bone disease (osteoporosis).

5. What is the Taeeumin (Taeeum type) like?

Taeeumin constitute 50% of the entire population.

1) External attributes (body)

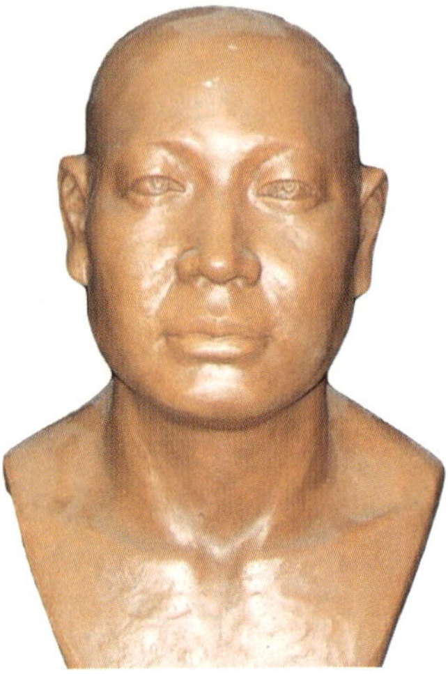

Taeeumin

The rounded facial features (specifically, the blunt nose, thick lips, and well developed jaws) and the typical moonface can confer a generous and dignified impression. They are careful speakers, voicing the few words they actually utter in low-pitched tones, which accentuates the treacherous and insidious nature typical of the Taeeumin. Their skin feels thick and rough to the touch, often covered in profuse sweat.

The Taeeumin usually stands in a composed posture, possibly owing much to their typical body build of well developed lower back and upper stomach and the poorly developed neck and hand. They are generally tall in height—short Taeeumin are rarely found. They are also typically overweight, and the few who manage keep trim are nonetheless sturdily built. Although abundantly endowed with brawn and flab, some Taeeumin can even be spotted in beauty pageants displaying the sensual figures they had carved secretly through rigorous exercise and diet.

As explained, the Taeeumin can easily be distinguished by their physical characteristics, but deciding on the constitutional type can sometimes prove to be more difficult than expected. It must be remembered that the Taeeumin can resemble the Soeumin closely in appearance in some cases.

2) Internal attributes (mind)

(1) Temperament and calibre

Consistent and calm, the Taeeumin have the capacity to persevere through any assignment once it's undertaken and to see through it to the end. They also have a natural ken to discern diligence and indolence in a man.

Taeeumin are highly adaptive people who easily settle down anywhere, well-known for their consistency and conservative tendencies that refuse to give in to pressure.

(2) Natural disposition

The innermost mind of the Taeeumin is by nature diffident and timorous, a state that arises from "wishing to attain amply what was sought after." When this penchant is left uncurbed, the state of mind deteriorates into a craven and pusillanimous one. At this moment the Taeeumin is too overcome by fear to start out on a new endeavor and in the end loses initiative. He repudiates all forms of change and retreats further into his secluded word. If things change for worse, the Taeeumin develops palpitations of the heart, a sign of adverse pathological development in the Taeeumin.

(3) Personality

The outbound facet, which arises from the Seong or the intrinsic selfless psyche, of a Taeeumin personality is always seeking stability and resisting change.

The inbound facet, which arises from the Jeong or the extrinsic self-centered psyche, of a Taeeumin personality focuses on building up the "inside affairs," and hence refuses all forms of change that might intimidate this. They show little interest on anything beyond family and domestic affairs, and in this they are poles asunder from the Soyangin who concentrate on matters that are irrelevant to themselves. When this sentiment becomes excessive, the following state of overweening desire ensues.

(4) Overweening desire

The most conspicuous manifestation of the overweening desire in

Taeeumin is the avarice for material possession, which becomes more pronounced when the desire to stay unchanged heightens into one to hoard and amass. It is the Taeeumin's forte to be able to secure his personal life tightly, but avarice can spring out of this if mild attachment turns into possessive obsession.

As aforementioned, the Taeeumin easily convey an impression of honesty and generosity, probably owing much to the heavyset facial features and the careful manner of speech and behavior. They are trustworthy people who act with honesty and prudence in every way possible, mostly conservative and wary of change, and always mindful of the politics of behavior. They are more often that not found to have a respectable social standing, for they are gifted with perseverance and determination, but the focus of their attention in life is the family and household, and they leave little interest or energy for the world outside. Although they convey an image of a mild-mannered gentleman, their hearts can carry mistrust and greed inside. Cowardly heart, sluggishness of action, and incorrigible laziness score as a few of the common faults found in the Taeeumin. Being of a sedentary and quiet nature, they typically prefer gambling over sports.

3) Symptomatic patterns

The following are the symptomatic patterns of the physiological state, non-critical pathological state, and critical pathological state found in Taeeumin.

Intact sweating mechanism is the key indicator of healthiness in Taeeumin. The least activity may leave a healthy Taeeumin perspiring profusely; a plate of warm food on a wintry day will produce the same results. Unlike the Soeumin who break down upon shedding a couple droplets of sweat, Taeeumin are completely immune to such events, indeed finding it invigorating to sweat at times. Thus, a healthy Taeeumin can be recognized by his profuse sweating.

Palpitations of the heart is the most significant manifestation of the mon-critical pathological state found in Taeeumin, which is set off when diffidence turns into cowardice.

More severe forms of illness arises from intensified imbalance of the Seong and Jeong. When this takes place, there is a noticeable increase of laxity in character, and physically the Taeeumin develops dry skin, sometimes experiencing decreased sweat. At other times, he develops diarrhea, because of which the mid-lower burn-center, the system related to the small intestine, can become clogged, leading to stifled, congested sensations. These are critical pathological symptoms which obviously demand medical treatment.

Apart from those mentioned above, common disorders found in Taeeumin include cerebrovascular disease, circulatory disease, respiratory disease, intestinal disease, skin disease, etc.

6. What is the Soeumin (Soeum type) like?

Soeumin, who constitute about 30% of the whole population, are known for their characteristic lean figure and the slightly feminine disposition. They frequently suffer from digestive disorders and resemble in many aspects what the Westerners have labeled "a cold constitution."

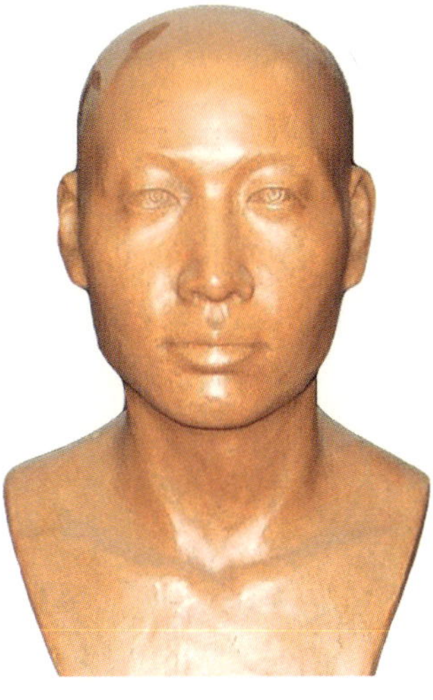

Soeumin

1) External attributes (body)

The delicate facial features and a slight antimongoloid slant of the eyes (downward slope of the eye aperture) together create a picture suggestive of beauty, for men and women alike. Their low-pitched, wispy, soft spoken voices are often employed in unexpected loquacity. They possess creamy white, soft and tender skin that is seldom covered in sweat.

Their body frame confers an especially stable sitting posture—stooped shoulders and constricted chest supported by a wide pelvic girdle and well developed hips, a proportion echoed in the four extremities, with the lower limbs considerably bulkier than their upper counterparts.

"Diminutive" and "slim" are the words most adequately descriptive of the typical Soeumin somatotype, which contrasts acutely with the sturdy Soyangin build. Owing to nutritional improvements and increased exercising, many Soeumin today resemble the Taeeumin in appearance, a point that demands careful observation in constitutional typing.

2) Internal attributes (mind)

(1) Temperament and calibre

Known for their clear-cut and composed character, the Soeumin's straightforwardness repudiates all forms of contortion, and yet this composure does not develop into boldness or pretension. A cautions decision-maker, the Soeumin often procrastinates at the crucial moment and lets a good chance slip by. Prudent in all matters and steeped in logical thinking, the Soeumin can be curt in interpersonal relationships. They have a knack for discerning the competence or incompetence in a man. They also tend to socialize only with those they are already familiar with, finding it an onerous task to cultivate new acquaintances.

(2) Nature disposition (Hang-shim)

In the deepest corner of the Soeumin mind exists a sentiment of insecurity and uncertainty, which arises from "wishing to attain everytime what was sought after." The digestive system, which according to the theory is the organ system most likely to go awry in the Soeumin, remains functional when this insidious feeling of insecurity is kept in check. But once the imbalance of Seong and Jeong is exacerbated, the Soeumin becomes excessively meek and compliant, physically developing gastrointestinal disorders accompanied by incessant diarrhea, in turn causing icy cold sensations in the nether-navel regions.

(3) Personality

The outbound facet, which arises from the Seong or the intrinsic selfless psyche, of the Soeumin personality is focused on the self. The Soeumin finds his faculties too preoccupied and overloaded with his own affairs to be able to extend and take interest in those around him.

The inbound facet, which arises from the Jeong or the extrinsic self-centered psyche, of the Soeumin personality always inclines to the "feminine" characteristics, being primly elegant, refined, and composed, sometimes demure and modest. When this penchant is amplified, the following state of overweening desire ensues.

(4) Overweening desire

The overweening desire of the Soeumin is expressed in the form of envy and furtiveness, the result of an excessively conceited mind and the craving to take away from others.

As already explained, the Soeumin possess a lean and petite figure that together with the delicate and refined facial contours conjures a feminine impression. They are also of a matching personality—meticulous and cautious, but also introverted and overreliant on logical thinking, making them poor socializers. The most prominent pathological traits found in the Soeumin are the frequent digestive disorders and the characteristic signs of a "cold constitution."

3) Symptomatic patterns

Described in the following paragraphs are the symptomatic patterns seen in physiological state, non-critical pathological state, and critical pathological state.

The absence of digestive problems is the most representational sign of unimpaired health in Soeumin. In Korea, Soeumin are often described as "having a cold constitution," which means showing the characteristics of cold extremities, hypomenorrhea (too little volume of menstrual bleeding), and increased vaginal discharge—all descriptions of the typical Soeumin. According to the Korean Medicine theory, a chronic deterioration of the digestive system is the fundamental cause conducive to these symptoms. Therefore, treatment of the digestive system and appropriate managing would be required for an expeditious recovery.

When the uncertainty and unsecurity amplifies, the Soeumin progresses into a pathological state that is fortunately non-critical and does not require particular treatment. Here, the Soeumin can experience a form of throat disease in which a hard, obstructive phlegm stuck in the throat can refuse to be expectorated or swallowed.

A more severe and critical pathological condition can arise when the imbalance of the Seong and Jeong worsens. On such occasions which demand medical interventions, the Soeumin turns exceedingly meek and compliant in temperament, physically developing incessant diarrhea and icy coldness in the subumbilical regions.

Some other forms of Soeumin pathology include digestive disorders, allergic disease (rhinitis, asthma, dermatitis), chronic fatigue, cold sensations (in extremities and abdomen), loss of weight, etc.

Archives: Frequently asked questions and misunderstandings about constitutional type

The following are excerpts from "Ask us Anything" that aired on KBS (Korean Broadcasting System), May 21, 2002, 10:00-10:45. The two guests appearing on the show are Lee Eui-ju (Professor of Korean Medicine at Kyung Hee University) and Kim Hyeong-gyu (Professor of Internal Medicine at Korea University).

1. What is 'constitutional type ?'

Host 1: Professors, I understand you two work in completely different fields. Is "constitutional type" something that actually exists?

Kim: Every person has his fair share of idiosyncrasies, and different people can show different reactions to the same disease or circumstances. If that's how we define "constitutional type," then a person's constitutional makeup does have significant meanings.

Lee: Constitutional typology is already an authenticated diagnostic tool in Korean Medicine (KM). In medical practice, you can recognize certain traits and patterns shared by a certain group of people, and I think if this can be fit into an organized scheme, that's where the fundamentals of constitutional typology can be found.

Host 2: Then what do you think constitution is? How would you define it?

Lee: Constitutional type could be defined as "the aggregate of physical form and functional character that is shaped by the hereditary determinants and environmental factors." The definition of constitution can differ slightly from person to person depending on the main field of study. Constitutional typology is a part of the medical tradition that has been received with much interest and research in countries both Eastern and Western, to this day.

Kim: The word 'constitutional type' is officially not a part of the modern medical terminology. You could say that constitution type can be understood as the characteristics of a person that is shaped through genetic and environmental factors, or a person's idiosyncratic reactions to external forces. The term is used in a very broad sense nowadays, but I think it can be succinctly stated as the trait differences between individuals.

2. Frequently asked questions and misunderstandings about "constitutional type"

Panelist: Most people seem to believe that constitutional typology is a concept found exclusively in Korean Medicine. But I've heard that constitutional theories had existed in the Western world as well.

Kim: Medical traditions around the world shared much in common until the 16th century, when Western medical practitioners discovered specific pathogens responsible for each disease for the first time in history. The concept of constitution centers around the stipulation that each individual's idiosyncrasies can be gauged up to a certain level. This was the conventional thinking, and modern medicine has since then deviated far from that perspective.

Host 1: I understand that in Korean Medicine, "constitutional type" is a concept that generally pertains to the Sasang Constitutional Typology?

Lee: In the Eastern world, the constitutional typology starts off in 『Internal Classics』, the archetypal tome in Eastern medicine, which mentions "the five types of people (五態人)." The Sasang Constitutional Medicine now applied in common medical practice was first introduced in 『Dongeui Soose Bowon』 by Lee Je-ma in the Joseon dynasty. Lee Je-ma observed four constitutional types (Taeyangin, Soyangin, Taeeumin, Soeumin), each with unique physical and mental qualities reflective of the underlying physiology and pathology found exclusively in that constitutional type, which then demands treatment and preservative care specific to that constitutional type. The branch of Korean Medicine studying this constitutional typology is called Sasang Constitutional Medicine. There are three main criteria involved in the typing process: external appearance, internal psyche, and symptomatic pattern. All the criteria must be analyzed together since each taken separately will not yield a comprehensive assessment. To be more precise, the symptomatic pattern criterion would come foremost among the three in significance. Constitutional typing is not something that can be done through mere observation of the facial feature, body frame, or personality trait. The most critical elements are the pathology and therapeutic reactions particular to that type of constitution.

Host 2: If we were to type constitution in modern medicine, what could the criteria be?

Kim: The two key determinants would be the genetic factor and the immunological factor. The hereditary factor can be assessed through genetic analysis, while the immunological factor is related to the psychological or environmental settings.

1) Family members of direct lineage all share the same constitutional type?

Host 2: Well, more people than you'd think believe that family members of direct lineage are bound to share the same constitutional type. Is this in fact true?

Kim: Well, certainly, there probably would be similarities, but not always. Constitution is largely dependent on genetic factors, and in most cases, your constitutional makeup would have much in common with your parents.' But when it comes to 'grandfather-father-son' line, there would be only 20% similarity. Another point to keep in mind is the "acquired" familial traits that can develop when generations of the same family are exposed to the same environment.

2) Experiences like delivering a child or serving in the military can change the constitutional type?

Host 2: Well, I myself have experienced some changes in body functions after giving birth to my child, like change in appetite, for example. Would you say that child delivery and military service could possibly have some bearing on your constitutional state?

Lee: Of course, there are variations between individuals, but pregnancy and childbirth are, in a way, a process of growing another organism inside your body. Naturally, the body goes through a cataclysmic transformation that can disrupt the internal balance and the biological flow of bodily functions. So it could be said that such event does have a great bearing on your constitution. Indeed, it significantly influences the mother's internal mechanisms. Likewise, living is strictly controlled in the military, meaning that the external environment surrounding the body is drastically changed, which is bound to influence the constitution. However, such changes result in shifts in the health levels of the body rather than the actual alterations in the constitutional type.

3) It's possible to classify the constitutional type by your blood type?

Host: Indeed, there is a lot of speculation on how the blood type can contribute to your constitutional type. Some even say that hematological typing is the modern day medical version of constitutional typing. What would you say to that?

Kim: There are over 60 blood types discovered to date, including the more famous ABO blood group and the Rh blood group. If we are to accept the blood type as a personal trait, it could be stated that there are in fact 60 constitutional types. However, constitution is not something that can be determined using a single standard, and it would be incorrect to state that blood typing systems can contribute to constitutional typing. What is granted in modern medicine is that the individual traits can all be accepted as personal particularities, but blood types are not enough to encompass the complexities of constitutional typing.

Lee: I have encountered people who believe in blood group-based constitutional typing, not just among layman, but also among practitioners as well who apply this in actual medical practice. But the blood type and the constitutional type are two completely different systems. All blood types, including the ABO types, can coexist in a single constitutional type, and vice versa. People tend to connect blood types with personality traits. They'd say, because persons with type O blood are 'universal donors' who can conversely receive only type O blood, they are bound to be equally altruistic in temperament. And because persons with type AB blood are 'universal recipients' who can only donate to other type AB blood recipients, people presume they are egoistic in temperament. But the ABO blood typing system is merely a simple classification of the hematotype into four types depending on the coagulation patterns found in a serum test. Naturally, it doesn't represent personality traits, and it is certainly unrelated to the Sasang Constitution Typology system. It's okay as long as people take it the way they take daily horoscope, but it does not carry any more significance than that.

4) Some constitutional types are more susceptible to certain kinds of illnesses?

Host: Some people catch cold or the flu all the time, would that have any pertinence to their constitutional types?

Kim: There are times when we feel particularly weak or react more sensitively to different factors, and this can undermine our immunological systems. I suppose that's the rationale behind the idea that constitution has something to do with the diseases we get. It's not a completely correct notion, but I have to admit that there can be a partial connection.

Panelist: A lot of people talk about balancing the body pH level. They also say that an excessively acidic condition makes you more vulnerable to diseases than high-alkaline constitutions. Is that really true?

Lee: Terms like "acidic body" or "alkaline body" are in themselves inappropriate. They apparently attempt to describe the acidity of the body, but the body pH is a fluctuating quality dependent on numerous variables, and acidic condition does not immediately entail susceptibility to disease. Staying healthy is not that difficult—the crux is to eat well, stay active, and live a regular life—but maintaining homeostasis is not so simple. Certainly, there are less chances of discovering alkaline-induced diseases than acid-induced ones, but still there are plenty out there who can attest that one can become sick even in alkaline states, and it cannot be assumed that "acidic body" immediately entails a heightened susceptibility to disease.

5) Drugs can change the constitutional type?

Host: Drugs that claim to improve constitution are pouring into the market. Do you think medication can actually influence the constitutional type?

Kim: Gene therapy products are not yet in common use, nor on the market, but anti-allergy drugs and immunosuppressive drugs that can regulate the immune system are currently being prescribed in daily medical practice. When immunological disorders are restored, the symptoms may subside temporarily, creating the impression of "improved constitution." However, immunological diseases are rarely completely cured, and the best that modern medicine can do is to merely improve or alleviate the symptomatic manifestations of the underlying pathological conditions.

Panelist: Some say that Korean medication can improve the constitution, something that cannot be achieved with Western drugs?

Kim: Whether Korean or Western, medication alone is not enough to alter the constitutional makeup. The main difference is that Western drugs are largely synthetic and refined whereas Korean medicaments are wholly natural. It's less a matter of superiority between medical divisions than the patient's preferences.

6) Is the O-ring test-based constitutional typing authentic?

Host 2: The O-ring test is carried out by forming an O-shaped ring with the thumb and index and then testing the muscular strength to check the individual's compatibility with an object, food, drug, etc. But most people experiment with this diagnostic tool just for fun, which I think surely undermines the authenticity of the test?

Lee: The official nomenclature for this testing is the "Bi-digital O-ring test," also known as the Omura test, after its inventor. It is a common procedure in Europe, the Scandinavian countries, the US, Japan, and other places scattered around the world. The O-ring test is theoretically rooted in the study of cerebral functions, and since the modern medicine does not yet have a comprehensive understanding of the brain functions, the underlying methodology in the Omura test is bound to contain shortcomings. Yet, the significance of this test lies in that it introduces an innovative and objective method of measuring qi (氣). As in applied kinesiology, the O-ring test measures the decrease of muscular strength the participant experiences when in touch with an incompatible object. Unfortunately, the reliability of this test is reduced by the abundance of variables that can affect the results and validity of the test, such as the side of hand being test, the directions faced while being tested, or the examiner's skills. Moreover, there are not enough theoretical validations to allow the O-ring test as a meaningful diagnostic tool in constitutional typing.

7) Can fasting therapy or raw-vegan diet change the constitution?

Panelist: Indeed, I've seen a lot of people who fast for days on or take up raw-vegan diet, purportedly to change their constitution.

Host 2: Truly, there are so many exaggerations about the effects of fasting or eating raw-vegan. People actually fall for it too, but then, there are others who claim to have actually benefited from them, and it's hard to know what to believe.

Kim: Fasting puts the body under a great deal of pressure and exerts a drastic change in the metabolism, and in this manner, it can be said that it does influence the constitution. And raw-vegan diet can be seen as a form of shock therapy that galvanizes the internal balance of the body by following a restricted diet. But indiscriminate fasting or extreme and unbalanced diet can prove to be hazardous for your health, bringing more loss than gain. The change in external environment brought on by the extreme diet could possibly affect your constitutional state in small scales, but it will be absurd to apply a uniform therapy to everyone without considering the individual variations. Plus, one should never forget that the diets can bring harmful results as well as beneficial ones.

Host: So, you're saying that changes in diet can actually influence your constitution?

Kim: Dietary habit plays an important role in maintaining health. By adjusting your eating habits, you can improve not only your constitution but also your general heath level, fortifying the immune system and strengthening the digestive system.

8) There is an eating style most compatible and specific to each constitutional type?

Host 2: I sometimes feel that my body "clicks" with certain kinds of food or eating habits better than others, a feeling anyone can experience by trying out different foods for themselves. I've heard a lot of traditional medical practitioners tell that certain foods can be good or bad for a certain constitutional type. What would you say to that?

Lee: What they are saying is largely based on Sasang Constitutional Medicine. Naturally, there can be restrictions in eating for certain diseases or certain constitutional types. For Western and Eastern medicine alike, there are certain kinds of food unsuitable for specific diseases, and others more adaptable for certain constitutional types. It is not merely a matter concerned with the type of constitution. The body's reaction to food depends heavily on the individual traits, and this subject needs to be tackled with a more comprehensive perspective.

Host: Well, to conclude, it can be said that the important thing is to take a more comprehensive view and make a balanced assessment of the present physical conditions to find the most compatible eating style. Well, then let's talk about our next topic.

9) We will be able to change our constitution with future advancements in genetic technology?

Panelist: If all the mysteries of genetics are uncovered, wouldn't it be possible to make the "perfect human being" by combining the more evolved genes?

Kim: It is the prevalent idea that there are specific genes responsible for corresponding diseases. Also, hereditary factors are known to be the dominant determinant for a person's constitutional frame, the aggregate of personal traits and characteristics. Considering these points, it is possible to speculate that once the human genome project is complete, it will naturally be feasible to genetically engineer a DNA sequence coding a new, perfected trait. However, there are ethical issues attached to this that need to be sorted out before any of this can be materialized.

10) There are differences in constitution between men and women, and among regions and races?

Host: Sometimes, I can't help believing that there is an unfathomable chasm between men and women, which I guess accounts for the ceaseless quarrels all couples can attest for. As they say, "men are from Mars, women are from Venus."

Lee: According to traditional medical theories, difference in constitutional types arises from the diversity in the internal organ structure and individual personality traits, rather than from gender, region, or chronological time period. Still, the incidences of each of the constitutional types can vary among groups or peoples at times.

Panelist: What does modern medicine have to say about this?

Kim: Men and women are different on the genetic level, specifically in the chromosome structure itself. Among different races, hereditary factors inevitably play an important role, since different races also have different genetic pools. Environmental factors are also significant, considering that races tend to stay in a particular habitat for generations; regional differences entail environmental differences.

Host: Why do you think people are so keen on getting to know their constitutional type? Why do they attach so much meaning to it?

Kim: Perhaps because it's the simplest way of explaining individuality, or it could possible have something to do with the psychological security one feels when he finds himself classified into a category that explains the background reasons for his personal traits. The security forged by the act of labeling is a significant factor, as is the satisfaction of being able to make deductions on individual attributes based on the personal constitutional type.

11) On constitutional improvement

Host 2: Most people interested in constitutional theories focus on constitutional improvement tactics. Could you possibly summarize the current perspectives and known facts on constitutional improvement?

Lee: Constitutional improvement is the process of regulating the physical state of the body from one of unhealthiness or illness into one of healthiness. Its ultimate goal lies in attaining the healthy and wholesome state (完實無病), and the essential and indispensable element (保命之主) that supplies the most basic life force, specific for each constitutional type. For example, we could discuss the constitutional improvement guidelines for the Soeumin: The first goal would be to treat digestive disturbances and thus achieve the state of "the healthy and wholesome state"; the second would be to attain "the essential and indispensable element," or the energetic life activity of the organs and related system, which in the Soeumin pertains to the "warm energy." As you can see, 'constitutional improvement' in Traditional Korean Medicine does not imply a transition into another type of constitution, but rather the regulation of the level of healthiness.

Kim: As I've said earlier, the concept of 'constitution' is more or less extinct in modern medicine. Still, the individual traits developed with the influence of the genetic and environmental factors could be seen as 'constitution.' We believe that the cardinal determinants for individual variation lies in the genes, one of the reasons that adds to the importance of studying genetics. Equally important is the environmental factor, closely related to the habitat and surroundings, which is one of the most significant contributors to healthy living, with or without bearings on constitution. In a broader perspective, we could say that all medical procedures and trials in the end ultimately aim for the betterment of health through improved constitutional type.

Chapter 3

The healthy way to live for each type of constitution

1. Constitutional type and food

There's a saying in Korea that food and medicine are of the same root, meaning that depending on the physical state, food can exert either a salubrious or pernicious effect on the body. This probably is the underlying rationale that accounts for the abundance of health food products found in today's society.

However, as many might have already experienced, an extreme or unbalanced diet can bring negative results. So to speak, insisting on a diet consisting of only the purportedly healthy foods can conversely be unhealthy for you.

A diet based on all of the food groups would be sufficient for most of us, whereas ill people would need further medical treatment other than dietary health care and would be served better with restrictive diet consisting of healthy foods. Only, since each constitutional type has its unique "Principals of life preservation,"[1] or the essential conditions for preserving vitality, there is also bound to be a regimen most compatible and specific to each constitutional type.

Principals of life preservation of the Taeyangin lies in the "absorptive and concentrative energy" (related to the deceleration of the qi and fluid metabolism[2] and the storage of energy), and thence the foods that support this absorptive and concentrative energy are beneficial (foods that can typically facilitate urination).

- **Beneficial food:** buckwheat, non-glutinous millet, non-glutinous rice, walnut, crab, oyster, common octopus, octopus, crucian carp, shrimp, turban shell, cuttlefish, abalone, sea cucumber, sea mussel, nori (*Porphyra tenera*), persimmon, kiwi, crimson glory vine (*Vitis coignetiae*), chinese quince, banana, cherry, orange, pineapple, grape, lettuce, chinese cabbage, celery, watershield, tomato, clear strained rice wine, coffee, root bark of araliaceous shrubs, grape root, pine needle, etc.
- **Harmful food:** strong or rich food, meat, wheat products, etc.

Principals of life preservation of the Soyangin lies in the "yin and cool energy" (related to the clear and cold energy of the water and nutrient metabolism),[3] and thence the foods that sup-

---

1 Principals of life preservation
The essential stipulations for preserving vitality.

2 The qi and fluid metabolism
The physical energy of the body is created through the qi and fluid metabolism and the water and nutrient metabolism. The lung and the liver are the two main components of the qi and fluid metabolism. The lung emits qi from the inside to the outside to be dispersed, whereas the liver absorbs qi from the outside to the inside to be stored. The lung produces energy and sweat, whereas the liver produces blood and urine.

3 The water and nutrient metabolism
The physical energy of the body is created through the qi and fluid metabolism and

port this yin and cool energy are beneficial (foods that are typically of a cold nature and able to facilitate bowel movements).

- **Beneficial food:** non-glutinous millet, barley, maize, barnyard millet, mung bean, red bean, sesame seed, frog, goose flesh, pork, hare meat, egg, duck egg, snakehead, flatfish, crayfish, crab, oyster, common octopus, sessile tunicate, globefish, shrimp, freshwater snail, carp, soft shell turtle, ear shell, sardine, herring, sea cucumber, strawberry, melon, banana, mountain berry, watermelon, oriental sweet melon (*Cucumis melo var. makuwa*), pineapple, apricot, eggplant, korean lettuce, chinese cabbage, lettuce, celery, purslane, sponge gourd, curled mallow, young radish, cucumber, burdock, bamboo sprout, plantain (whiteman's foot), pumpkin, green tea, draft beer, cassia tora seed, chinese matrimony vine fruit, steamed rehmannia root (chinese foxglove), aloe, red reishi mushroom, frozen desserts, safflower, bay salt (sun-dried salt), safflower oil, etc.

Principals of life preservation of the Taeeumin lies in the "emissive and dispersive energy" (related to the acceleration of the qi and fluid metabolism and the dissipation of energy), and thence the foods that support this emissive and dispersive energy are beneficial (foods that can typically facilitate sweating).

- **Beneficial food:** millet, yam, wheat, wheat flour, bread, indian millet, maize, adlay, brown rice, potato, starch, sweet potato, taro, kidney bean (haricot), soybean, tofu (soybean curd), hazelnut, perilla seed, peanut, chestnut, ginkgo nut,

---

the water and nutrient metabolism. The spleen (related to the "warm energy") and kidney (related to the "cold energy") are the two main components of the water and nutrient metabolism. The spleen elevates energy from the lower part of the body to the upper part and digests food. The kidney lowers energy from the upper part of the body to the lower part and passes stool. The spleen generates qi and participates in digestion, whereas the kidney produces essence and participates in defecation.

pine nut, walnut, beef, butter, milk, cheese, hairtail (cutlass fish), crab, mackerel, whale meat, saury (mackerel pike), codfish, pollack roe, brown croaker, japanese eel, japanese icefish, turban shell, trout, salmon, freshwater snail, carp, yellow corvina, tuna, nori (*Porphyra tenera*), laminaria, brown seaweed, agar, jellyfish, japanese apricot, pear, apricot seed, mulberry, plum, bracken, carrot, bonnet bellflower, balloonflower, japanese angelica tree shoot, butterbur (bog rhubarb), radish, east indian lotus root, korean daisy (*Aster koraiensis*), soybean sprout, pumpkin, oak mushroom, oyster mushroom, ear mushroom, sangwhang mushroom (*Phellinus linteus*), pine mushroom (*Tricholoma matsutake*), winter mushroom, soybean paste (doenjang), sugar, soybean milk, soju (distilled liquor), yoghurt, velvet deer antler, squalene, schizandra fruit (*Schizandra chinensis Baill.*), longan, liver oil, thick beef soup, chrysanthemum, evening primrose oil (sundrops), beef soup, arrowroot, etc.

Principals of life preservation of the Soeumin lies in the "yang and warm energy" (related to the clear and warm energy of the water and nutrient metabolism), and thence the foods that support this yang and warm energy are beneficial (foods that are typically of a warm nature and able to facilitate digestion).

- **Beneficial food:** non-glutinous rice, maize, hulled millet, glutinous millet, glutinous rice, whole wheat flour, potato, starch, hazelnut, dog meat, pheasant, roe deer meat, chicken, quail, snake, mutton, goat meat, sparrow, turkey, hare meat, flatfish, hairtail (cutlass fish), sand lance, sea bass, codfish, sailfin sandfish, red seabream, catfish, anchovy, alaska pollack, chinese fine scaled weather loach (*Misgurnus mizolepis*), clubbed tunicate, brown croaker, japanese eel, japanese icefish, flathead mullet, freshwater mandarin fish, ayu sweetfish, southern conger eel, yellow corvina, sea mussel, laminaria, mandarine, jujube, lemon, peach, apple, pomegranate, yuzu (*Citrus junos Tanaka*), brown mustard, red pepper, shepherd's purse, wild rocambole, carrot, garlic, dropwort, chinese chive, ginger, spinach, japanese mugwort, crown daisy (garland chrysanthemum), cabbage, onion, tomato, welsh onion, mustard, salt, curry, pepper, unrefined rice wine, sheep milk, licorice root, cinnamon bark, korean angelica, royal jelly, northern groundcorn, chinese motherwort, ginseng, siberian motherwort, bittern, leaf beet, honey, malt, lacquer plant, etc.

(For further reference, see Appendix 3.)

2. Eating habits

The foods we eat nurture us and provide us with the energy to stay active, think, see, hear, and taste.

An improper eating habit is something not easily broken, and most of us aren't even able to recognize a bad eating pattern when we see one. Nevertheless, one's eating habits, a manifestation of the person's innermost nature, can easily influence health, and pernicious eating habits that can undermine our health must be corrected immediately.

Interestingly, each type of constitution manifests a unique and specific eating style that reflects the innate nature and temperament of that constitutional type.

The typical Taeyangin is either fastidious about his food or completely indifferent. He eschews rich or strong food, preferring instead cold dishes or plain-tasting food. Meat, dairy products, and wheat products can exacerbate indigestion, a condition that commonly stems from the underlying gastroesophageal disturbances, and should be avoided.

The typical Soyangin gorges on his food, sweeping up whatever food on the table. Blessed with a solid digestive system, he is rarely met with severe digestive conditions. Yet, the hasty eating habit sometimes induces digestive disorders, and a longer mealtime spent savoring the dishes is required. He also tends to frequently skip meals and binge later—a habit largely responsible for the Soyangin indigestion. In order to preserve health, the Soyangin must cultivate a regular and moderate eating habit.

The typical Taeeumin is a gourmand with a huge stomach, eating under the motto "quantity over quality." He finds abundance the standard for a satisfactory meal and is unable to abstain from food, often stuffing himself to the neck. Intolerant of hunger, he can never leave a crumb on the plate. He typically favors sweet and rich food, eats quickly and in large amounts, and habitually forgoes breakfast and compensates later in the day with a sumptuous dinner. However, the calories taken in are often not burned with matching physical activity, which inevitably leads to obesity. The Taeeumin is required to regulate his food intake in order to live a healthy life.

The typical Soeumin eats light, valuing quality over quantity. Blighted with a weak digestive system, he is a gourmet with a small and fastidious appetite. A nighttime snack or inadvertent overeating can induce dyspepsia in the Soeumin. They are choosy and picky eaters and thus often find themselves in malnourished and underweight conditions. Hence, it is recommendable for the Soeumin to eat small yet balanced meals and exercise afterwards to help digestion.

3. Ways of self-expression for each type of constitution

"It's all in the presentation," one has said, stressing the importance of self-expression tactics for surviving the modern society; hence the myriad of Korean mothers who impose extracurricular self-expression tutoring lessons on their underaged pre-school kids.

Self-expression is a process of communicating the inner thoughts and emotions to the outside world through words, gesticulations, and facial expressions. Though it depends on small scales on the person's domestic environment and education, the innate self-expression skills are rarely improved dramatically through teaching.

By examining and analyzing the self-expression tactics and the concomitant results found commonly in each type of constitution, we can make our social life run more smoothly and pleasantly.

The typical Taeyangin is an eloquent speaker who can captivate the crowd and lead them on, sometimes presenting brilliant and innovative ideas. On the other hand, he can be overly self-righteous and is poor at accepting criticism, rejecting any contention or casting aspersions to those who challenge his ideas.

The typical Soyangin is a witty and humorous speaker who displays fluent conversation skills and uses abundant body gestures to convey his meaning, traits that confer persuasiveness to his speech. On the other hand, the typical loquacity can give an impression of garrulity, and his speeches are filled with slips of the tongue, while his inconsiderate and impetuous remarks may offend many of his listeners.

The typical Taeeumin is taciturn and reticent, often reluctant to express himself. They do not waver in their opinions to the end and enjoy lecturing and preaching. Sometimes, the audience are overtaken by the sheer presence and air of dignity that exudes from the humongous body frame typical of the Taeeumin. On the other hand, his parlance is often garbled, disorganized, and illogical, which can naturally frustrate his listeners.

The typical Soeumin is a systematic and logical speaker. His parlance or gestures are not grandiloquent, but the words spoken are orderly and calmly persuasive. On the other hand, he tends to take criticisms to his heart, even though he is careful to hide his grievances from others.

4. Image make-over

Admittedly, everyone wants to be good-looking. The modern day society is an extremely competitive one, and to a certain extent one can understand the reasons for the sudden popularity of plastic surgeries, especially afore a job interview.

According to a recent survey asking bachelors in their twenties on plastic surgery, more than 70% of the respondents replied that confidence in his appearances is related to a positive and active career, and that they feel positive about their own girlfriends' plastic surgery. Also, a Korean woman invests 58 minutes a day in grooming, and looks into the mirror 5 times a day, on average. Apparently, it is a common desire, for both men and women, to cultivate a pulchritudinous image of oneself.

And yet, first impressions are not solely determined by the appearances. Generally, a person's image is a combined reflection of his appearances and impression. Considering this, what one can do is to realize the how he appears and what kind of impression he gives to others, and then to cultivate the given looks into a desirable image.

The typical Taeyangin has a charismatic air that can sometimes be interpreted as intimidating, and he is also apt to give an acerbic and agitated impression. Frequent smiles and less eccentric behaviour could perhaps soften this image.

The typical Soyangin has pointed eyes, a sharp nose, and an oval face, altogether giving a frosty first impression that makes him difficult to be accosted. One solution to this could be to emphasize his humorous and witty qualities and to cultivate a cheerful and energetic image.

The typical Taeeumin gives a brusque and headstrong impression. This could be remanded by accentuating the characteristic artless and steady qualities found in the Taeeumin, or in the case of a woman, by assuming the "generous matron" attitude.

The typical Soeumin can appear absent-minded, with glimpses of slyness underneath. His refined and polished appearances can impress favorably, but the typical aloofness and haughty attitude can put people off. In order to make himself more approachable, the Soeumin is required to communicate and interact with others more actively with a broader perspective on life.

5. Dressing style

It is said that a woman looking at a man first sees around the neck—the necktie and the shirt—and moves on to observe how good he looks in his clothes; it is said that a man looking at a woman first takes in her silhouette. Clearly, one's dressing style has a highly visible quality and thus becomes the primary criterion in judging a person's first impression.

The dressing style is a sort of ornamentation that can show what the person is like by reflecting his social rank and position or by expressing his individuality. And so, it would be wise of us to improve our dressing styles by matching them with the dressing mode most compatible to our constitutional types.

The typical Taeyangin prefers to wear retro. Insensitive to trend, he dislikes giving change to his dressing style and chooses practicality over vogue. On the other hand, he sometimes enjoys dressing in antique clothing, a peculiar taste that attests to his originality.

The typical Soyangin prefers an extravagant and shocking style. He is one of the people who believe themselves to be trendsetters, opting for bold styles without thinking much about the onlooking eyes. He favors intense hues of chroma, and if he finds an outfit unbecoming, he refuses to wear it ever again. Even in the most frigid of winters, he characteristically chooses being fashionable and cold to being covered thick and warm.

The typical Taeeumin rarely pays much attention to fashion and rather prefers wearing comfortable outfits. His most exciting colors are at best plain, and he is completely ignorant when it comes to fashion trends. He occasionally attempts a bold deviation from his usual drabness but instead often ends up looking awkward.

The typical Soeumin is usually well tuned in to the latest trend, but still sticks to an unexciting dress code. In his world of fashion, buttons are meant to be buttoned to the last buttonhole, and a meticulously combed hair is a must-have for a day's serenity of mind. His primary value lies in a well-matched appearance, and more often than not, the color scheme of his outfit, hair style, shoes, and accessories all fit together in a completely natural and uniform flow. Perhaps because the Soeumin has a weak spot for cold weathers, his outfit tends to precede a season too early.

6. Marital life

One of the major social issues during the IMF monitored economic crisis in Korea was the disintegration of families.

According to a statistical data on 'divorce by consent' in post-economic crisis Korea, the age distribution among the divorced ranged from twenties to sixties, most commonly citing economic issues and irreconcilable differences as the reason for the split.

What was most surprising was that whereas most divorces in the past were between couples in their twenties or thirties citing irreconcilable differences, the more recent divorces were often between couples in their fifties or sixties citing economic issues and irreconcilable differences.

In the conventional Korean family culture, spouses would have drawn closer and consoled each other during crisis to overcome the difficulties together. What the deteriorated modern day family culture lacks and needs is marital support between spouses based on knowledge and understanding of constitutional differences between partners.

For example, though Taeyangin-Taeyangin couples are rare, they need to respect each other's individuality. Taeyangin are generally passionate, but can sometimes be violent and self-righteous and downright loathsome, at other times dreamers who seek for the classic romantic, platonic love.

To maintain a good relationship, Taeyangin-Taeyangin couples need to stop interfering in each other's personal life and avoid extreme mood swings. It can be a good idea to accompany spouses to, say, elegant suburban music cafes, or take a trip to an out-of-the-world wilderness attraction, or listen to some music together, or go fishing for a change from time to time.

Soyangin-Soyangin couples are short-tempered and extroverted, with an extravagant and a la mode taste in dressing. They share a common lack of interest in domestic duties that are often pushed aside for other self-gratifying activities. Naturally, they constantly plunge into incessant squabbles that are fortunately also short-lived. Soyangin are renowned for their assiduity, an attribute that can boost their marital life into a most interesting one, with the help of a little tolerance and leniency from both sides.

Soeumin-Soeumin couples yearn for a calm and tranquil domestic environment. They are loath to deal hands-on with conjugal conflicts. They have a strong drive to keep their private lives separate, and a breach of the personal boundaries or any show of disregard or disrespect for the partner can seriously marr affections for quite long.

Taeeumin-Taeeumin couples rarely have conjugal conflicts owing much to their bovine temperaments. However, the relationship can seem pretty monotonous because they are both unanimated and undesirous of excitement. On one side, they are persistent and obstinate, and on the other, they are willing to be the first to bend in a quarrel, and an apologetic word is enough to melt away hostilities. They unexpectedly manage to maintain a quiet tone of love throughout their marriage by popping up with hearttouching surprise gifts from time to time.

Taeeumin-Soeumin couples may appear to be in a loving relationship to the outsiders' eyes, but the truth under the veneer of affection in many cases is that they have already given up on each other. A happy conjugal life can be guaranteed only if they strive to understand each other's shortcomings—the Taeeumin's uncouthness and the Soeumin's prissiness. What the Taeeumin often forgets and should never forget is that generally the Taeumin's fondness for the Soeumin is in fact intense and real.

The Soyangin almost always takes up the role of the troublemaker in a Taeeumin-Soyangin relationship. Fights spark off in the most absurd fashion, but the Taeeumin is generally magnanimous, and the Soyangin usually apologizes first, not being one for bearing grudge, and makes up for it with the characteristic Soyangin brand of charm, which can sweep away any resentments held against him.

Many Soeumin-Soyangin couples find their fantasies of each other shattered after marriage. They feel more and more alienated from each other as they discover more and more differences. This needs to be stopped before the situations get out of hand and cause them to become estranged; it's important to keep the old flame burning by acting like newlyweds throughout married life.

The essential understanding of the Sasang Constitutional Medicine lies in "knowing and doing." Lee Je-ma proposed "understanding others," and "rectifying oneself," as the two primary goals to be attained in order to perfect knowing and doing.

In the end, the same rules apply to conjugal harmony in marital life. Here, understanding others would lie in showing respect for the partner's constitutional attributes and in allowing the partner to live accordingly, and rectifying oneself would lie in recognizing one's own fallibilities and in compensating for them.

7. Taking care in spring

The four seasons and the day-night rotations allow the growth, development, and fructification of all life on Earth. In other words, the Yang energy that is strengthened in spring and summer encourages the growth and development of living things, while the Yin energy that is strengthened in autumn and winter encourages the fructification and culmination of living things.

In spring, the energy that had been condensed during winter is released into all directions. In a way, it is similar to how the potential energy in a nuclear bomb explodes and spreads outward, taking the form of a great mushroom cloud.

If this galvanizing and revivifying energy of springtime is not conducted and released well in the human body, it consequently results in drowsiness, lassitude, inertia, etc. Measures are hence necessary to facilitate the adaptation process that passes in a body in order to acclimate to the seasonal changes. We will examine the more common springtime maladies and the preventive measures that can be adopted for each constitutional type.

First, eye diseases are common in spring, especially the irritant conjunctivitis, allergic conjunctivitis, keratitis, and corneal ulcer. Sandy dusts from the dry parts of China and Mongol is the main cause of the irritant and allergic conjunctivitis. The pollutants and heavy metals mixed in the dusts, such as iron, silicon, copper, lead, cadmium, and aluminium, affect the eye and cause roughness and irritation. Rubbing the eye can inflict ruptures on the cornea that can lead to bacterial infection, resulting in keratitis. Birch, alder, and oak pollen, which are commonplace in spring, cause the allergic conjunctivitis. Irritation, sneezing, epiphora (watering eye), and chemosis (swollen eye) are some of the main symptoms of allergic conjunctivitis, more frequently found in Taeyangin or Soyangin. The most effective preventive measure would be to stay indoors as much as possible and to avoid the pathogens responsible for the inflammation. Unavoidable outdoor excursions could be made safer by wearing hats and sunglasses to protect the eye and by washing hands as often as possible to fend off possible pathogen invasions. Rubbing the eye is strictly discouraged, and itchiness or irritation should be treated by rinsing the eye with saline solution or lukewarm water.

Second, many people catch cold in spring due to an increase in the diurnal temperature range. The typical Soeumin cold characteristically starts with rhinorrhea (nasal discharge) and sneezing. It is sometimes accompanied by digestive disorders or diarrhea. Medicaments capable of producing light sweating can be applied in such conditions, such as Pogostemonis Herba, Cnidii Rhizoma, Perillae Folium, Allii Radix, and Cinnamomi Ramulus. Some of the other beneficial foods include ginger tea, dried pollack soup, apple, and mandarine. It is imperative that the light sweat induced by the medicine does not develop into profuse sweating.

The typical Soyangin cold characteristically starts with severe body aches, thirst, and tonsillitis. It is sometimes accompanied by constriction in the chest or urinary problems. Medicaments that can be applied include Ledebouriellae Radix, Notopterygii Rhizoma, Schizonepetae Herba, and Bupleuri Radix, and some of the other beneficial foods include seafood soup, cabbage soup, and strawberry. Soyangin disorders tend to progress rapidly, demanding early interventions.

The typical Taeeumin cold characteristically starts with hypohidrosis (sweat deficiency), inertia, tension in the head and neck, thirst, and laryngopharyngitis. It can sometimes be accompanied, if the hypohidrosis remains untreated, by a bad case of enteritis. Medicaments that can be applied include Ephedrae Herba, Puerariae Radix, Cimicifugae Rhizoma, and Armeniacae Amarum Semen, and some of the other beneficial foods include radish soup, bean sprout soup, soy pulp (okara), and pear. An appropriate approach to the Taeeumin cold would be to apply medicational therapy in the early stages to generate profuse perspiration.

The typical Taeyangin is seldom afflicted with colds. Even when he happens to have acquired one, he usually complains of lower back pain and oliguria (lower than normal volume of urine) rather than the quintessential symptoms like rhinorrhea, sneezing, or diarrhea. Medicaments that can be applied include Acanthopanacis Cortex, Chaenomelis Fructus, kiwi, and grape. The typical Taeyangin cold generally passes lightly, and an excessive and superfluous use of medicine is hence strictly forbidden.

Last, the spring fever is another malady found commonly in spring. The symptoms of spring fever range from sensations of simple fatigue to unconstrainable drowsiness in daytime. The solution to this offered in Korean Medicine is sufficient sleep at night, adequate exercise, and balanced nutritional absorption. The weak-constitutioned Soeumin should especially be more wary of malnutrition, while the heavy-set Taeeumin is urged to exercise and sweat off the typical inactivity. The Soyangin or Taeyangin are recommended to take a brief daytime nap or to supplement themselves with vitamin.

According to 『Neijing』, a famous classic work in Internal Classic Korean Medicine, "Spring is the time when all living things, of the sky and of the earth, are revitalized and energized. The man will sleep late and wake up early, walk the gardens, wear garments light and comfortable, keep the mind generous and refrain from appropriating or punishing. Thus is the way of the spring regimen. Follow these rules, and the energy of liver will acclimate to the energy of spring and remain intact." To our ancestors, a life harmonious to nature was regarded the paragon of life, and they understood the ways of conforming to the seasonal flow. However, modern day people disregard the seasonal changes and its significance, consequently suffering from the resulting illnesses and diseases. Wisdom is required to live harmoniously.

8. Adequate spring food for each constitutional type

The fresh spring herbs can spice up lost appetites and shake off inactivity. The lush spring harvest can be used to fend off the inertia and lassitude associated with a typical spring fever. Spring foods, combined with the sufficient knowledge of their effects and compatibility for each constitutional type, can benefit our health immensely.

Let us examine the effects of some spring food on our health.

Food type	Taeyangin	Taeeumin	Soyangin	Soeumin
Crab	○		○	
Leaf beet	○	○	○	○
Shepherd's purse	○	○	○	○
Wild rocambole	○			○
Japanese angelica (fatsia)	○	○	○	○
Strawberry			○	
Chinese weatherfish (mud loach)		○		○
Water dropwort				○
Spinach	○	○		
Japanese mugwort				○
Garland chrysanthemum				○
Lettuce (Ixeris dentata)	○		○	
Curled mallow			○	
Korean daisy				

(○: constitutional type that benefits from the foodstuff)

(1) Shepherd's purse (pickpurse)

The pickpurse is neither too cold nor too warm in nature, has a sweet taste, and is high in protein, making it ideal for those with a weak digestive system. It is especially effective in treating menstrual disorder, epistaxis (nose bleed), and postpartum hemorrhage caused by physical weakness. It can produce favourable effects on aged people, or on those suffering from chronic fatigue due to hepatic dysfunction, or even on hangovers. It is not restricted to any particular constitutional type.

(2) Wild rocambole

The wild rocambole is warm in nature and has a spicy taste, an effective herb for treating insomnia and in increasing stamina. It fortifies the blood system according to traditional herbalogy, and is thus often included in medicaments for gastritis, metrorrhagia (uterine bleeding), or paramenia (abnormality of menstruation).

It is abundant in vitamin C, and helps combat skin aging. It strengthens the hepatic system and prevents arteriosclerosis. However, being of a warm nature, it can produce unpropitious effects when taken in large doses by those with a "hot constitution," meaning having a lot of heat in their system. It is most beneficial for Soeumin and Taeyangin.

(3) Fatsia shoots

The fresh fragrance that emanates from a fatsia shoot sprout soothes the mind, a quality that confers its tranquilizing and sedative effects, applicable for the people who are especially listless in the mornings or for those living under constant stress, like business workers and students.

It strengthens kidney functions for those afflicted with nephric insufficiency, chronic kidney disease, or the concomitant edema and urinary frequency. It is recommended for all constitutional types.

(4) Dandelion

A commonly found spring herb, the dandelion is either temperate or slightly cool in nature. It is thus effective in extinguishing excessive body heat and exhibits anti-inflammatory actions, features that mark it as a preferred drug for various inflammatory diseases. It is used for boils in various parts of the body, halitosis caused by physical weakness, cystitis, and vaginitis. It is recommended for all constitutional types.

(5) Lettuce (Ixeris dentata)

The Ixeris dentata plant is a herb that closely resembles the lettuce, rarely found in the West but abundant in Korea. Its coldness of nature is conducive to its efficacy in eliminating heat and unhealthy energy from the internal organs. An outstanding herb for the spring fever, it relaxes the mind and body and drives away drowsiness.

Some of the applications of this herb include mastitis, severe cough, xerostomia (dryness of mouth) combined with loss of appetite, and erythruria (redness of urine) or urethral discomfort. It is recommended for Soyangin and Taeyangin.

9. Surviving the summers

As the ancient saying goes, a scorching summer and a frigid winter give good harvests. We often experience a feeling of debility caused by excessive sweating in the summers. Sweat can be defined as bodily fluid composed of water and electrolyte. Traditional Korean Medicine conceives sweat as 'Sweat = Energy + Fluid'; hence, the more the sweat, the more the exhaustion of energy and fluid. In Korean tradition, there are "canicular days" on which people feast on special nourishing soups (boshintang, samgyetang) in order to replenish the lost energy. Wisdom is required to survive the summers healthily.

According to a report on the relationship between vacations and heart diseases among American workers, those who work without vacations suffer a twice as high incidence of heart disease compared to those who take all the vacations offered to them. This implicates that summer vacations can have significant bearings on the mental and physical well-being of an average worker.

1) Taeyangin

(1) Summer healthcare

A Taeyangin is not easily swayed by the summer heat, but he needs to keep stress under control, be it work-related stress or interpersonal stress. Another point to keep in mind during the summer is to prevent oliguria (decrease in urine volume) by eating simple and plain-tasting foods and absorbing sufficient fluid.

The typical Taeyangin has either an especially fastidious palate or a completely indifferent one. He generally dislikes rich or strong food, preferring vegetables and plain food. Beneficial foods for a healthy summer include plain-tasting food, buckwheat, octopus, sea mussel, shellfish, chinese quince, kiwi, grape, orange, etc. Harmful foods include strong and heavy food (meat), dairy product, wheat product, alcoholic beverage, etc.

A Taeyangin should avoid exhaustive exercising and choose activities that can train both the mind and body, such and qi-gong, martial arts (wushu), or swordcraft. Power work-outs (high intensity work-outs during a short time span) and cardiovascular training (ex. 100m dash, short-distance cycling, 50m swimming, etc.) are some of the recommendable exercises. The Taeyangin has a well developed back and a poorly developed abdomen, suggesting stretching and flexibility exercises for the head and back and sit-ups for the abdominal muscles.

(2) The perfect vacation spot

A Taeyangin is more often than not gifted in the arts, displaying talent in music, fine arts, calligraphy, etc., sometimes delving deeper by actually playing the instruments or producing artistic works. We urge the Taeyangin to drop in on one of the art performances in the summer—the artistic absorption will blow away the summer heat. Or how about learning an instrument—the mellifluous notes, after enough time and practice, will stir up new dreams in the heart.

Some Taeyangin have a bent for history, philosophy, politics, etc. For these people, we suggest a dig in the library or browsing the internet in search of the cities or regions related to a specific topic, followed by an actual visit to the museums and historical sites. Be prepared with a thorough preliminary research on the subject—the background knowledge of the historical facts will make the trip more savory and worthwhile.

The more ambitious ones aspiring to land in the higher ranks of the business world can prepare for the high-fizz world by reading useful books on the subject—lofty dreams will swallow up the summer swelter.

2) Soyangin

(1) Summer healthcare

There is a famous saying in Korea, "Eat hot to ward off heat," but the Soyangin is an exception to this rule. A typical Soyangin is thin and prone to the disorders related to fire and heat in traditional medicine, such as constipation. To spend a healthy summer, it would be wiser of him to maintain a tranquil mood and enjoy the summer fruits.

The typical Soyangin gorges on his food, sweeping up the food on the table in the blink of an eye. Despite this pernicious eating habit, he is immune from serious digestive conditions, being blessed with a solid digestive system. Yet, the hasty eating habit is conducive to rare but severe attacks of digestive disturbances, and the eating habit needs to be broken by training for a different eating pattern, such as eating more slowly in extended mealtimes, taking ample time to savor the rich taste of the food on the plate. Beneficial foods for a healthier summer include barley, watermelon, oriental sweet melon, strawberry, oyster, lettuce, chinese cabbage, sea cucumber, abalone (ear shell), loach soup, and pork. Harmful foods include honey, garlic, red pepper, and dog meat.

A Soyangin is certainly not the patient type to follow through with the original physical training plans. Rather than setting unattainable goals, he should start out with easier and steadier routines that span out over three months, with three exercise sessions each week. We recommend resistance training (anaerobics) such as weight-lifting, wrestling, and bar exercises. The Soyangin has a well developed chest and a poorly developed lower body (lower back and hip), suggesting stretching and flexibility exercises for the chest muscles and muscle building exercises for the lower part of the body.

(2) The perfect vacation spot

Go find a wilderness attraction. Hands-on experiences will add to the thrill and excitement.

Paragliding, rafting, mountain biking, or horseback riding are some of the activities a Soyangin might enjoy. The thrill and excitement discovered in the out-of-the-world wilderness, shared with friends, will be more than enough to make a memorable summer. Riverside water-skiing and yachting could be another solution for the torrid summer days.

Some may find a family trip to the local amusement park or theme park appealing, while still others may opt for a holiday package stay in a luxurious hotel fully accoutred with commodious swimming pools and sauna.

Those interested in the fashion world could satiate the typical Soyangin appetite for extravagance with a visit to a fashion show. Be prepared to be consumed by the dazzling catwalk on the runway, the splendor of the haute couture, the meticulously arranged exhibition on the glamorous stage. And those who find their heart in the film industry could easily hop to the cinema and be purged of the summer heat, showered and saturated with the magnificent artistry on the silver screen.

3) Taeeumin

(1) Summer healthcare

Most Taeeumin, who typically sweat more than any other type of constitution, dislike the summer season. However, with the adequate survival tactics they will find the summer less of the usual bore. For the Taeeumin, we recommend a leisurely sojourn in the nearby mountains or seaside, combined with extra care to provide the body with enough fluid to make up for the water loss.

The typical Taeeumin is a gourmand with a particularly large stomach for food, eating under the motto "quantity over quality." For him, a large meal always qualifies as a good meal. He is unable to stay away from food and often catches himself over-eating. Beneficial summer foods for the Taeeumin include thick beef soup, adlay, ginkgo nut, bonnet bellflower, bean, beef, and milk. Harmful summer foods include coffee, chicken, and pork.

Taeeumin are generally averse to activity and by nature sedentary, more often found immersed in computer games. We prescribe them a diversity of exercises that involve much activity and equally profuse sweating. Exercises that develop cardiovascular endurance are recommended, namely mountain hiking, marathon, jogging, long-distance swimming, long-distance cycling, etc. The typical Taeeumin somatotype is characterized by well developed abdomen and poorly developed head and back, suggesting stretching and flexibility exercises for the abdomen and muscle building exercises for the head and back.

(2) The perfect vacation spot

Anywhere near water will do for the Taeeumin.

More than anywhere else, an island will prove to be worthy of a visit, especially if one can sail out on a dinghy and go fishing in the sea, to discover the unexpected joys of a hands-on angling experience. An impromptu seafood party spiced with the right booze will provide instant gratification. Drop off at the beach and indulge in the excitement of water activities or simply recline and wallow in the sun.

A day spent at the luxurious spa retreat will be perfect for an elderly Taeeumin couple. Sweating from the torrid summer sun, the source of infinite annoyance for the Taeeumin, is one thing, sweating off in the spa is another. The exhilaration felt at the end of a good sweating in the hot spa waters is a treat given by god exclusively to the Taeeumin.

If the summer break was just a step short of fully quelling the heat, a visit to the neighborhood swimming pool could be another option—a choice that serves both as an escape from the scorching summer and a boost for a healthier lifestyle.

4) Soeumin

(1) Summer healthcare

The Soeumin is prone to summer colds from too much air conditioning and bouts of digestive disorders due to the typically weak digestive system. To avoid an unpleasant summer, the Soeumin is required to abstain from indulging in summer fruits or cold beverages and to exercise regularly to stay fit.

The typical Soeumin eats light, valuing quality over quantity. Blighted with a weak digestive system, he is more of a gourmet type person, with a small and fastidious appetite. His picky eating habits often result in malnutrition and being underweight. Hence, it is recommended that the Soeumin eat a small but balanced meal and exercise afterwards to facilitate digestion. Beneficial summer foods for the Soeumin include nourishing soups (samgyetang, boshintang), glutinous rice, glutinous millet, jujube fruit, chinese chive, ginger, curled mallow, japanese mugwort, and chicken. Harmful summer foods for the Soeumin include beer, buckwheat, barley, oriental sweet melon, and chestnut.

Soeumin, unlike the other types of constitution, are one of the odd people who perversely tend to build up fatigue through exercising instead of being relieved of it. Work-outs that involve sweating are definitely on their pernicious activities list. What we can recommend for the Soeumin are physical activities that require stability of the lower body, say golf or baseball, that can take advantage of their sturdy lower back and lower limbs. They would find muscular endurance training most suitable, such as pull-ups, sit-ups, push-ups, free-exercise, or dumbbell exercises. They typically have well developed hips and poorly developed chests, suggesting stretching and flexibility exercises for the waist, hips, and legs and muscle building exercises for the chest.

(2) The perfect vacation spot

Find a quiet corner where a natural breeze can cool the summer heat. Remember, air conditioners, fans, and frozen desserts are the Soeumin's worst enemies.

Hidden in the midst of the numerous mountains and hills lie the log cabins—welcome to the forest campground. The time spent in this out-of-the-world place, practicing the pristine life-style, will make the tangled world seem like a remote memory of the past. Self-cooked meals will go easy on the stomach, while the cool wind in the valley blows in the face to swish off the sweat, the way no manmade air conditioners can.

Or take a book and sit in the shade. The fleeting wings of imagination will fly you to the uncharted worlds—where further could one possibly go?

The followers of the Buddhist teachings could journey to a Buddhist culture site or a Buddhist temple. Christians may desire to make a pilgrimage to the sacred places and meditate on the great love of Jesus. Those interested in Confucianism would do well to visit a nearby oriental classics learning center and cultivate the mind and body according to the ancient teachings.

All parts of Korea each have a distinctive speciality food to boast of. The regional attitude is often succinctly reflected in its foods. The gourmet in the Soeumin would hardly pass up the opportunities to delight, and soon be lost, in the kaleidoscope of flavors.

10. Constitutional type and sweating

We often experience a sense of debility caused by excessive sweating in the summer. Sweat consists of 99% water, and the remaining 1% is composed of NaCl (salt), KCl, lactic acid, urea, glucose, and other organic substances, all dissolved in the water in minuscule amounts.

Sweat can also be used as a diagnostic tool depending on the site of the perspiration on the body, sites that differ among the constitutional types. Thus, the sites of the 'healthy sweat' can be examined to determine the adequate amount, intensity, and time length of a physical performance (activity, labor, and exercise).

A Taeyangin needs to learn emotional control in order to resolve work-related stress or conflicts in the office. We also suggest that he avoid strong or rich food and abstain from alcoholic beverages. Oliguria can and should be prevented by choosing to eat plain-tasting foods and supplying himself with sufficient fluid. Sweating in the groin or lower back indicates an adequate amount of exercise or a 'healthy sweat.'

A typical Soyangin has a thin figure and is prone to the disorders related to fire and heat in traditional medicine, such as fatigue, weight loss, agitation, constipation, etc. It would therefore be wiser of him to maintain composure and let the sweet summer fruits cool off any insidious feelings of frustration. Sweating in the hands and feet indicates an adequate amount of exercise or a 'healthy sweat.'

A typical Taeeumin sweats profusely in hot weather, providing yet another reason to detest the season. But the summer will seem shorter with some splashing in the valley or on a beach, supplied with light and airy clothes and sufficient fluid to make up for the lost sweat. Swimming and other whole-body workouts will make him feel the temperature has suddenly dropped by several degrees. Sweating that starts in the forehead shortly after exercising and moves on to the face, neck, and chest indicates an adequate amount of exercise or a 'healthy sweat.'

A typical Soeumin is especially vulnerable to summer colds when exposed to excessive air conditioning. He has what Koreans conveniently term a "cold stomach," meaning having a weak digestive system that makes him prone to digestive disorders like diarrhea, stomachache, and dyspepsia, especially in the summer. He needs to avoid the summer fruits or cold beverages, and instead pamper the stomach with warm food. Excessive sweating exhausts energy, and it is recommended that the Soeumin exercise only lightly, just enough to improve his appetite and facilitate digestion. Sweating on the tip of the nose and the philtrum indicates an adequate amount of exercise or a 'healthy sweat.'

A sudden, unexplained, and significant increase of sweat should be suspected as a sign of an underlying medical condition. Cold sweat on the back suggests tuberculosis; sweat that leaves yellow stains on the underclothes indicates hepatic disease. Febrile diseases like hepatitis and typhoid fever induce profuse sweating, enough to soak the underclothes. Also, those in chronic infirmity must observe the urine volume carefully during a long-lasting heat wave.

Diabetics could undergo a sudden rise of blood sugar if an excessive amount of bodily fluid is lost through perspiration. The drainage of water from the body induced by sweating takes out the minerals along with it, leaving the system with lower levels of potassium, a valuable electrolyte that amplifies the effect of cardiac stimulants—a dangerous state in which the heart can fail to contract in response to drugs. Other than these, gastropathy cases like functional dyspepsia and chronic gastritis are also exposed to the dangers of electrolyte imbalance and fluid volume deficit caused by excessive sweating, a state that can exacerbate the gastropathy conditions.

11. Edible greenery

Sasang Constitutional Medicine asserts that only eating the most compatible foods for the constitutional type can potentiate the "Principals of life preservation," or the essential conditions for preserving vigour, and guarantee "State of healthiness," or the state of being free of pathological conditions. In other words, the essential elements for preserving vitality and the optimum state of health are specific for each type of constitution, which then demands health regimen specific for each constitutional type.

The popular belief that all vegetables are beneficial to health is an unsophisticated and puerile one compared to the ideology introduced in Sasang Constitutional Medicine. Foods are beneficial only when eaten with care by the most compatible type of constitution and can indeed exert pernicious effects when consumed by an incompatible constitutional type, and vegetables are certainly no exception to this rule.

For example, cucumber and oriental sweet melon, of cold nature according to traditional medicine, produce favorable effects on Soyangin, eg. by helping bowel movements, in contrast to the unfavorable effects on Soeumin, eg. by causing stomachache and diarrhea. This leaves us with the question, which vegetables are most salubrious for my type of constitution? To know this, you first have to establish your constitutional type and then proceed to discriminate the compatible vegetables based on the Sasang Constitutional Typology.

A typical Taeyangin dislikes rich and strong food, finding plain food and low fat dishes more appealing to his palate. He can thus eat any kind of green grocery and still maintain excellent health. Any vegetables, the likes of chinese cabbage and potato, are compatible, and plain tasting foods like buckwheat, persimmon, banana, tomato, chinese quince, wild rocambole, and grape are especially healthy. However, foods that can stress the digestive system, like wheat, and starchy, thick vegetable mixture products are harmful.

A typical Taeeumin is the antithesis of the Taeyangin in every way, including the choice of food. Foods purported to draw "heat" to the center of the body in the traditional medical view, like persimmon, chinese quince, kiwi, cherry, chinese cabbage, clam, and pine needle, should be avoided if possible. According to Sasang Constitutional Medicine, many Taeeumin are obese and binge habitually, just the type of person prone to cardiovascular diseases like hypertension, CVA, and heart disease, should enough "heat" accumulate in the body. Hence, foods that can dissipate and emit this "heat" are most beneficial, including root foods, such as radish, carrot, east indian lotus root, taro, yam, bonnet bellflower root, and burdock, and plain-tasting vegetables, such as balloonflower, bracken, pine mushroom, tofu (soybean curd), soybean sprout, soy pulp (okara), nori, and brown seaweed.

A typical Soyangin gorges on his food and eats quickly, a pernicious eating habit conducive to digestive disorders. Having the kind of constitution that the traditional medicine refers to as having a lot of "heat" in the system, the Soyangin are one of the surprising people who enjoy frozen foods and cold dishes even in the most frigid of winters. It is only natural that they are incompatible with the foods known to generate heat according to traditional medicine, such as ginseng and honey. Following the same logic, a jar of honey loaded with ginseng, something considered a special gift of honor in Korea, would bestow them less good than harm, causing constriction in the chest, or a sensation of rising heat, or even constipation. The foods that can actually benefit the Soyangin include the vegetables considered "cold" in traditional medicine, like cucumber, chinese cabbage, curled mallow, eggplant, pumpkin, burdock, lettuce, and water dropwort. Some grains are also beneficial, such as barley, red bean, mung bean, sesame seed, and non-glutinous millet, as well as certain fruits, like watermelon, oriental sweet melon, and strawberry. In contrast, the vegetables considered "warm" in traditional medicine, like garlic and red pepper, should be eschewed.

The typical Soeumin is known for his fastidious appetite and frequent bouts of indigestion, a quality that matches the description of having what is considered a "cold" constitution in traditional medicine. And naturally, the foods known to be easily digested and "warm" in nature are most compatible with the Soeumin, namely shepherd's purse (pickpurse), water dropwort, spinach, cabbage, crown daisy (garland chrysanthemum), potato, welsh onion, garlic, pepper, ginger, and chinese chive. But even these are considered to have traces of "cool" energy that needs to be driven out by cooking the material over heat, a rule that is also applied to the herbs used in a Soeumin herbal medication. Foods considered "cold" in nature and difficult to digest are forbidden, including buckwheat, chinese cabbage, pear, watermelon, cucumber, sweet potato, chestnut, walnut, mung bean, barley, red bean, and milk.

12. Taming of the mind (I): Psychological structures of the Sasangin

It seems that most cultures have a phrase pertaining to the "stomach-churning jealousy," epitomizing the ubiquity of the greedy human nature that covets for the material wealth and feats of another.

As we all know, controlling and correcting our own selves is the most difficult task in life. The mind and body of the self need to be trained and cultivated in order to rectify the self (a process termed "rectifying the self") before we can begin to comprehend others (a process termed "understanding the others").

Sasang Constitutional Medicine emphasizes this procedure of rectifying the self and understanding the others, a process of training the mind and body to eliminate covetous greed.

Lee Je-ma explained the structure of the human mind as having two compartments, one revolving around the self and one concerned with the non-self, or other people. The psyche concerned with the non-self, or others, gives rise to the natural disposition, and the psyche revolving around the self gives rise to the covetous greed. Therefore, understanding how the mind operates for each constitutional type will allow us to understand the self and further know about others comprehensively.

We will first examine the outbound compartment of the mind (the psychological reaction to others).

It is the Taeyangin's propensity to move forward without stepping back. This natural disposition allows him to associate smoothly in interpersonal relationships, along with the ability to read the good or evil in a man.

The Taeyangin is never free from a feeling of urgency. If this feeling is amplified, the Taeyangin may develop psychopathological conditions associated with insomnia and paranoia. As a remedy, the Taeyangin can expand his scholarly pursuits and accrue more knowledge, and thus train the mind to have more integrity and sincerity.

It is the Soyangin's propensity to shed light on a matter and raise interest on it. This natural disposition allows him to display decorum and moderation in interpersonal relationships, along with the ability to read the knowledgeability and ignorance in a man.

The Soyangin is never free from a feeling of apprehension. If this feeling is amplified, the Soyangin may develop psychopathological conditions associated with severe forgetfulness. As a remedy, the Soyangin can cultivate moderation and dignity, and thus train the mind to have more wisdom.

It is the Taeeumin's propensity to desire to remain constant without changing. This natural disposition allows him to excel in instructing and leading in interpersonal relationships, along with the ability to read the diligence and indolence in a man.

The Taeeumin is never free from a feeling of wariness. If this feeling is amplified, the Taeeumin may develop psychopathological conditions associated with palpitation, or throbbing of the heart. As a remedy, the Taeeumin can culture a more heightened perception of the world through a strategic outlook, and thus train the mind to have more industry.

It is the Soeumin's propensity to abide in a fixed place or position. This natural disposition allows him to be appeasing and soothing in interpersonal relationships, along with the ability to read the capability and incapability in a man.

The Soeumin is never free from a feeling of instability. If this feeling is amplified, the Soeumin may develop psychopathological conditions associated with inertia of the extremities. As a remedy, the Soeumin can improve his abilities and skills, and thus train the mind to have more capability.

We will now discuss the inbound compartment of the mind (the psychological structure of the self).

The Taeyangin displays active "masculinity" and conversely passive "femininity." This masculine inclination, if excessive, can develop into jealous envy of others' wisdom and sagacity. This masculine attribute acts advantageously in distant and impersonal relationships but disadvantageously in intimate and personal relationships, thus giving rise to flippant and self-indulgent characteristics. To overcome this, the Taeyangin needs to focus on cultivating intimate relationships.

The Soyangin concentrates on outer achievements and neglects inner security. This extroverted propensity, if excessive, can develop into jealous envy of others' business and career. This extroverted attribute acts advantageously on achieving in external affairs but disadvantageously on procuring internal security, thus giving rise to narrowminded and self-opinionated characteristics. To overcome this, the Soyangin needs to pay more attention to internal affairs.

The Taeeumin concentrates on inner security and neglects outer achievements. This introverted propensity, if excessive, can develop into jealous envy of others' material possessions. This introverted attribute acts advantageously on procuring internal security but disadvantageously on achieving in external affairs, thus giving rise to greedy and covetous characteristics. To overcome this, the Taeeumin needs to pay more attention to external affairs.

The Soeumin displays active "femininity" and conversely passive "masculinity." This feminine inclination, if excessive, can develop into jealous envy of others' abilities and skills. This feminine attribute acts advantageously in intimate and personal relationships but disadvantageously in distant and impersonal relationships, thus giving rise to furtive and guileful characteristics. To overcome this, the Soeumin needs to focus on associating with new acquaintances.

13. Taming of the mind (II): The vices

We are all observant in one way or another of a purpose in life, and we all find ourselves at one point or another in a meditative and introspective mood. As they say, even the great minds can discover a merit in the low minds to be learned of.

What sets apart the great minds from the low minds, the virtuous from the unworthy, the illustrious from the lowly? Most human vices revolve around the four temptations: alcohol, sex, money, and power.[4] Breaking away from the temptations will

---

4 The vices

The vices are introduced in explanation of the ways to resolve conflicts arising in human life, or "Humanly activity." Lee Je-ma expounds that there are four aspects of interpersonal relationships: rapport between spouses, rapport between senior and junior, rapport between lord and servant, and rapport between parent and offspring. He also put forth the basic elements in the relationships: "abidance," "assemblage," "association," and "judgment," respectively. Defect of "abidance" in a relationship between spouses will marr the rapport and render it susceptible to the temptations of sex, defect of "assemblage" in a relationship between senior and junior will marr the rapport and render it susceptible to the temptations of alcohol, defect of "association" in a relationship between lord and servant will marr the rapport and render it susceptible to the temptations of power, and defect of "judgment" in a relationship between parent and offspring will marr the rapport and render it susceptible to the

confer the mind tranquility and the body rectitude, assimilating the state of the highest virtue in the oriental culture—the state without desires.

The Taeyangin needs to be vigilant of the temptations of alcohol. His fortes lie in "judgment" and his faults in "assemblage." Adroitness in "judgment" allows him to thrive in "association." However, he is maladroit in serving his seniors (those senior in age or experience) and coping with his junior, leading to his weakness in "assemblage." This is conducive to the Taeyangin's indolence and vulnerability to alcohol.

Being self-righteous and dogmatic, he is easily tempted to take up an isolated life. But relying on alcohol is far from the wisest way of stepping out of it, which would be to adopt a grander perception of the purpose of life and culture a more innovative mind that can fully unravel his potential.

In order to ascend to greater virtue, it is crucial that the Taeyangin maintain diligence of the body: keep distance from drinking, avoid idleness of the body, abstain from indulging in food, and cultivate the skills of "assemblage" in the relationship between senior and junior.

The Soyangin needs to be vigilant of the temptations of sex, or carnal desires. His fortes lie in "association" and his faults in "abidance." Adroitness in "association" allows him to thrive in "judgment." However, he is maladroit in tending to his spouse, leading to his weakness in "abidance." This is conducive to the Soyangin's licentious extravagance and vulnerability to sex.

Impulsive sexual drives must be curbed, for both men and women. Such urges can also appear with underlying medical conditions, in which case can be discriminated distinctively from physiological desires.

In order to ascend to greater virtue, it is crucial that the Soyangin maintain simplicity and austerity of lifestyle: stay distant from carnal desire, refrain from extravagance of ornamentation,

---

temptations of money.

build mutual trust between husband and wife, and cultivate the skills of "abidance" in the relationship between spouses.

The Taeeumin needs to be vigilant of the temptations of money, or material goods. His fortes lie in "assemblage" and his faults in "judgment." Adroitness in "assemblage" allows him to thrive in "abidance." However, he is maladroit in a parent-offspring relationship, leading to his weakness in "judgment." This is conducive to the Taeeumin's greed for material gain and vulnerability to money.

Fiery greed is similar to riding out into the endless sea. The Taeeumin needs to learn the wisdom of being satisfied with the present and instead helping other friends in need.

In order to ascend to greater virtue, it is crucial that the Taeumin expand his experiences: purge the mind of material greed, divert away from single-minded pursuit of material gain, cultivate mutual affinity between father and son through guidance, and focus on developing the skills of "judgment" in the relationship between parent and offspring.

The Soeumin needs to be vigilant of the temptations of power. His fortes lie in "abidance" and his faults in "association." Adroitness in "abidance" allows him to thrive in "assemblage." However, he is maladroit in a lord-servant relationship, most similar to the employer-employee relationship today, leading to his weakness in "association." This is conducive to the Soeumin's feelings of insecurity and vulnerability to power.

He is inclined to group into exclusive cliques that refuse to accept the majority viewpoint. We suggest to the Soeumin: be wiser, share opinions with the crowd, and accept the merits of others with grace.

In order to ascend to greater virtue, it is crucial that the Soeumin be guarded against bigotry: purge the mind of the greed for power, step away from the manipulations and subterfuge in the scramble for yet higher and loftier position, cultivate smoother relations at workplace, and develop improved skills of "association" in employer-employee relationships.

Archives: Knowing your own constitutional type can help you survive the summer

The following are excerpts from "Ask Us Anything" aired on Korean Broadcasting System (KBS), July 24, 2001, 10:00-10:40. The guests appearing on the show are Lee Eui-ju, the author of this book, and Prof. Jo.

1. Summer health and the constitutional type

Host 1: Now, let's talk about summer health guidelines for the different constitutional types. Before we begin, what are the most common summer-related illnesses for each constitutional type?

Lee: The Taeyangin and Soyangin find it difficult to stand the heat, whereas the Taeeumin and Soeumin usually complain of summer colds from having too much air conditioning. The "yang" types often manifest the symptoms related to what is conventionally known as "being affected by the heat," including darkened facial shades, high body temperature, constriction in the chest, or dry mouth. Conversely, the "Yin" types show the symptoms related to summer colds that comes from having too much air conditioning or eating too much cold foods, manifesting the quintessential symptoms of the common cold such as headaches, runny nose, sneezing spells, often accompanied by gastroenteric symptoms like diarrhea, dyspepsia, and cold extremities.

2. The appropriate summer exercise for each type of constitution

Panelist: It's really tempting to lay stretched out without doing anything in the summer. What kind of exercise would you say would be most appropriate for each constitutional type, and what are the things to watch out for?

Lee: I recommend lower body work-outs for the "yang" types and upper body work-outs for the "yin" types. More specifically, the Taeyangin would benefit from 100m dash, short-distance swimming, short-distance cycling; the Taeeumin would benefit from marathon, mountain climbing, jogging, long-distance swimming, long-distance cycling; the Soyangin would benefit from weightlifting, wrestling, gymnastics; the Soeumin would benefit from pull-ups, sit-ups, and push-ups. If you're a beginner, exercise moderately, just enough to sweat in the palms, 3-5 days a week, for 2-6 weeks. Intensify the work-outs over the next weeks—increase slightly the time and intensity every 2-4 weeks. If you're obese, the exercises can stress or injure the knees and ankles, so start at an easy pace and tone up systematically.

3. Appropriate summer foods and herbal medication for each type of constitution

Host 1: Whichever constitutional type you are, the summer is bound to be a difficult time of the year. You have to watch what you eat constantly to avoid exhausting out and to stay fit. Would there be any beneficial or harmful foods for the specific types of constitution?

Jo: The Taeyangin are inclined to suffer from digestive degeneration in the summers, a condition that can be alleviated through a careful diet of plain and low fat dishes. Conversely, high calorie food, high fat food, and heat producing foods are to be avoided. Some of the beneficial foods include root bark of araliaceous shrub, grape, kiwi, pine needle, and crucian carp. The Taeeumin can benefit most by following the traditional aphorism "eat hot to fight off heat." Blessed with a strong digestive system, they benefit most from high protein, high calorie meat products and other foods that fortify their weak "lung system," provided it is eaten in moderate amounts. The foods most suitable for the Soyangin include certain grains (barley, red bean, mung bean, buckwheat), meat (pork, duck), vegetables, fruits (watermelon, oriental sweet melon), and cold foods (frozen desserts). The Soeumin generally lose appetite in the summer and often suffer from bouts of diarrhea, a condition exacerbated by eating "cold" fruits and vegetables, like oriental sweet melon, watermelon, cucumber, etc. Avoid raw or cold foods and instead eat well cooked foods and nourishing recipes (samgyetang, chinese weather loach, etc.).

Panelist: We naturally prefer cold dishes when it's hot. Are there different limitations for different constitutional types concerning eating "cold" or "hot" food?

Lee: "Cold" foods like buckwheat, barley, or frozen desserts are recommended for the Taeyangin and Soyangin, but these can produce pernicious effects on the Soeumin, who are typically of a "cold" constitution. The Soeumin will instead benefit from "warm" and "hot" foods, which can conversely harm the Soyangin; the nourishing soups (samgyetang, sacheoltang) eaten on special occasions works like gasoline on fire on the Soyangin.

4. Revitalizing herbal medicine for the summer

Host 2: It is a common custom in Korea to take revitalizing herbal medicine to fight off the typical summer lassitude. But there are people who say that the medication is inefficacious because it is all discharged with the sweat. What would you say to that?

Jo: The metabolic rate increases in the summer and more waste material can consequently be discharged. This however does mean that the effects of the medicine is diminished. And not everyone has to take revitalizing medications in the summer. It is the weak and elderly or those in infirmity who should be prescribed with medicines to help survive the summer, when the physical functions tend to degenerate.

Chapter 4

Disease Management for Each Constitutional Type

1. Personal health care (I): Understanding "Principals of life preservation"

Anyone can be his own home doctor if he can recognize the indicators of health for each constitutional type. Our Korean ancestors applied practical wisdom in their households, a tradition now nearly lost.

Let us begin by examining the term "Principal of life preservation." In a narrow sense, the Bomyeongjiju is the main energy that maintains and preserves life; in a broader sense, it is the energy that supplements and supports the innate constitutional makeup and plays a role in the physiological or pathological developments that ensue.

The principal of life preservation has a complementary quality different for each constitutional type: the Taeyangin and Taeeumin Bomyeongjiju are each related to the storage or dispersion process in the qi and fluid metabolism, and the Soyangin and Soeumin Bomyeongjiju are each related to the "cold" or "hot" energy in the water and nutrient metabolism.

The principal of life preservation of the Taeyangin lies in the "absorptive and concentrative energy." The Taeyangin is innately endowed with a strong "emissive and dispersive energy," a quality responsible for his tendency to advance without retreating and his strong masculinity. Likewise, the pathological conditions found commonly in Taeyangin are related to excessive saliva and foaming in the mouth, reflux of food, etc. Therefore, the "absorptive and concentrative energy" that can counter this eruptive propensity becomes the essential element for constitutional health care in Taeyangin.

The principal of life preservation of the Soyangin lies in the "yin and cool energy." The Soyangin is innately endowed with a strong "hot energy," a quality responsible for his excitable and extroverted nature. Likewise, the pathological conditions found commonly in Soyangin are related to severe forgetfulness, constipation, etc. Therefore, the "yin and cool energy" that can counter this hot constitution becomes the essential element for constitutional health care in Soyangin.

The principal of life preservation of the Taeeumin lies in the "emissive and dispersive energy." The Taeeumin is innately endowed with a strong "absorptive and concentrative energy," a quality responsible for his reserved and introverted nature. Likewise, the pathological conditions found commonly in Taeeumin are related to palpitations of the heart, hypohidrosis (decrease in sweat), etc. Therefore, the "emissive and dispersive energy" that can counter this congestive propensity becomes the essential element for constitutional health care in Taeeumin.

The principal of life preservation of the Soeumin lies in the "yang and warm energy." The Soeumin is innately endowed with a strong "cold energy," a quality responsible for his strong introversion and femininity. Likewise, the pathological conditions found commonly in Soeumin are related to cold sweat, diarrhea, icy sensations in the lower belly, etc. Therefore, the "yang and warm energy" that can counter this cold constitution becomes the essential element for constitutional health care in Soeumin.

2. Personal health care (II): Understanding State of healthiness

We have previously examined the principal of life preservation for the different constitutional types. Let us now look into the indicators of health that are manifest in different forms for each constitutional type.

The secretions from our body consist of urine, stool, sweat, tears, and nasal discharge, which can be analyzed in order to gauge the state of healthiness, or the healthy state free from pathological conditions.

The state of healthiness of the Taeyangin depends on sufficient amount of and clear quality of urine. The urine is the end product of qi and fluid metabolism that reflects the state of the "absorptive and concentrative energy."

The state of healthiness of the Soyangin depends on an unconstipated bowel state. The stool is the end product of water and nutrient metabolism that reflects the state of the "yin and cool energy." If the bowel movements are stalled and become constipated, it implies a shortage of the "yin and cool energy" and the subsequent increase in the "hot and fiery energy."

The state of healthiness of the Taeeumin depends on free passage of sweat. The sweat is the end product of qi and fluid metabolism that reflects the state of the "emissive and dispersive energy." If perspiration of the body become caulked, it indicates an underlying pathological condition that is interfering with and concentrating the "emissive and dispersive energy."

The state of healthiness of the Soeumin depends on unimpaired digestion. Digestion is the end product of water and nutrient metabolism that reflects the state of the "yang and warm energy." Indigestion and diarrhea indicate a shortage of the "yang and warm energy" and the consequential increase in the cold and icy energy.

3. Winter colds and the constitutional type

'Why do people catch cold in the winter?' The winter is the season when the dry and cold climate sends all living things into storage and culmination. For example, the leaves of the plants desiccate in the autumn as all the energy in the leaves and stem is transferred to the roots. The energy thus stored in the roots remains dormant in the winter and sprouts out with new life in the spring. Similarly, during winter, humans store energy in the "roots" inside our body, or the internal organs, causing the outside, or the skin and the meridian system, to weaken and wither and hence become exposed to infections.

The common cold is a commonplace malady in the winter, the specific symptoms of which can vary among people. Most regard the cold as a triviality, but distinct variations among them can be observed when closely surveyed case by case.

The typical Taeyangin is rarely afflicted with colds. Even if he happens to have acquired one, he seldom displays the characteristic symptoms of the common cold, instead reporting idiosyncratic symptoms like lower back pain or oliguria (less than normal volume of urine). This condition can be ameliorated with certain medicaments, such as root bark of araliaceous shrub, chinese quince, kiwi, or grape. The Taeyangin cold can easily be cured with light treatment—indeed, doctors are warned against overtreating the condition.

The typical Soyangin cold exhibits the initial symptoms of severe body aches, thirst, and tonsillitis. In 2-3 days, it develops into constrictive sensations in the chest or urinary discomfort.

This condition can be ameliorated with certain medicaments, such as Ledebouriellae Radix, Notopterygii Rhizoma, Schizonepetae Herba, Bupleuri Radix, seafood soup, chinese cabbage soup, or strawberry. The typical Soyangin disorders tend to progress rapidly, and early interventions are thus crucial.

The typical Taeeumin cold starts with the initial symptoms of hypohidrosis (sweat deficiency), heaviness of the body, tension in the head and neck, thirst, and laryngopharyngitis. It develops into severe enteritis in 2-3 days if the hypohidrosis is left untreated. This condition can be ameliorated with certain medicaments, such as Ephedrae Herba, Puerariae Radix, Cimicifugae Rhizoma, Armeniacae Amarum Semen, radish soup, bean sprout soup, soy pulp (okara), or pear. Generating sweat by taking medicine in the early stages is the most effective way of treating the Taeeumin cold.

The typical Soeumin cold shows the initial symptoms of rhinorrhea (nasal discharge) and sneezing. It is sometimes accompanied by digestive disorders and diarrhea within 2-3 days. This condition can be restored by generating light sweat through ingesting certain herbs like Pogostemonis Herba, Cnidii Rhizoma, Perillae Folium, Allii Radix, and Cinnamomi Ramulus. Some of the other beneficial foods include ginger tea, dried pollack soup, apple, and mandarine. Producing excessive sweat is strictly forbidden.

We have here examined the main outlines of the common cold. If the symptoms do not abate, or show signs of further bronchial or gastroenteric complication, consult the nearest traditional medical practitioner.

4. Overcoming fatigue

People who complain of fatigue are often dismissed as a whiner, especially if young, and are left without opportunities to treat the condition properly.

Even when they finally decide to tackle their problem and visit a hospital for a medical check-up, no culpable factors, at least none that are physical, are to be found and thus no solution.

In these cases, though no visible impairments can be diagnosed, a factor that interferes with the normal metabolism certainly and definitely exists. In the Sasang Medicine paradigm, the pathological conditions arise from an excess or deficit of the four essential elements that constitute our body: "Spirit," "Qi," "Blood," and "Essence."[1]

The Taeyangin needs to store "Blood" and "Essence." Fatigue can occur if the "Blood" and "Essence" fail to be stored, or if violent outbursts of emotion or prodigal drinking interfere with the storage of the two elements. This demands restraint of temper and a more abstemious lifestyle of regular hours and simple food.

The Soyangin needs to preserve the clear energy of "Blood" and "Essence." The hot and fiery energy induced by overheated metabolism can cause fatigue and further lead to constipation. This is similar to driving 50km/h in low gear. To overcome such conditions and the related symptoms, the Soyangin needs to deal life with composure and maintain moderation and dignity at all times. He should find temporary respite from the stressful life by making a trip to the mountains or grasslands. Like everyone else, he should learn to reward himself with peaceful slumbers.

The Taeeumin needs to disperse "Spirit" and "Qi." In order to do this, the Taeeumin is required to overcome indolence and material greed, and to remain studious and assiduous. He should not rely solely on medicine to relieve fatigue. Regular physical

---

1 Spirit Qi Blood Essence
The physical energy of the body is created through the "qi and fluid metabolism" and the "water and nutrient metabolism." The lung and the liver, and the spleen and the kidney are the main components for each system of metabolism, in respective order. The main energy source is converted into different forms of energy that is utilized differently in the metabolic process of the four components: Spirit Qi Blood Essence, in respective order.

training can discharge the waste material clogged up in the system through sweat, urine, and stool.

The Soeumin needs to preserve the clear and warm energy of "Spirit" and "Qi." To do this, the Soeumin is required to keep the body warm and to eat readily digestible foods.

5. The war against obesity (I)

The alarming increase of obesity, a condition related to many chronic and degenerative diseases, has been met with great concern in medical circles of both East and West.

Obesity is a condition of the body in which the surplus energy is converted into excessive fat storage, depending largely on two factors: one, excessive intake of energy, and two, insufficient consumption of energy.

An accurate analysis of the responsible factors for the weight gain and designing corresponding weight control plans are indispensable for fighting against obesity. A psychological approach to the problem is also important.

Already, innumerable solutions to weight problems had been proffered, including strict diet plans, exercising, psychotherapy, and surgical methods, but what we are about to discuss here is an innovative method of approaching obesity based on the constitutional differences.

The Taeyangin is generally of a lean figure and rarely obese. However, some tend to indulge in the negative emotions, particularly anger, and choose to live a reclusive life, often using alcohol to flush out the troubles of the outside world off his mind, a habit conducive to obesity. To avoid gaining weight, the Taeyangin needs to realize that abstaining from alcohol and retaining a tranquil frame of mind comes foremost on his list of priorities. After that, suitable physical training can help the Taeyangin keep in shape.

The Soyangin, being of a "hot" constitution, or having much "heat" in the system, are mostly thin, but some are born with a sturdy and stocky somatotype, the type of body frame that often leads to obesity. Fortunately, their extroverted and active traits help prevent excessive storage of energy. The typical binge eating habits need to be replaced by exercising habits.

Most Taeeumin are overweight due to their gluttonous appetite (leading to excessive intake) and their typically conservative and lazy personality (leading to insufficient consumption). They incline toward emotional eating, demanding therapies to correct this habit. Some of the possible treatments and solutions include psychotherapy, more active social life (hobbies and leisure activities), and rigorous physical training.

The Soeumin are seldom obese due to their weak digestive systems (meaning a small intake of energy). Yet, rare cases of overweightness are caused by improper eating habits (irregular meals, high calorie snacks) or insufficient physical activity that owes largely to their introverted personality. The logical solution to the Soeumin obesity would therefore be to eat regularly, avoid snacks, and exercise regularly.

6. The war against obesity (II): Weight loss tips for different constitutional types

A tall figure, svelte curves, a shapely nose, and razor-sharp chin lines are now accepted as stipulations of beauty—the credo behind the diet frenzy that has now consumed the whole nation.

But beauty is not the sole issue involved in the battle against obesity. Obesity can lead to other devastating conditions, and keeping within adequate weight levels is necessary in order to maintain health.

However, another thing to keep in mind is that the problem of obesity is a minefield that will explode into nasty results if tampered without the accurate knowledge and information.

We will discuss here the common mistakes that need to be avoided for successful weight loss for each constitutional type.

The Taeyangin are generally lean and rarely obese. However, they tend to fall prey to alcoholic dependency because of their inability to control the explosive surging of anger and frustration they experience when thwarted in their paths, being incapable of accepting methods or accomplishments other than their own. Since habitual drinking can lead to obesity, the Taeyangin needs to drop the deteriorating drinking habit and step away from the occasional outbursts of anger by emptying the mind.

(1) Lifestyle

① The Taeyangin is unrelenting to ways or accomplishments other than their own.

⇒ Do not cling on to self-made ideas and listen to the advices coming from more knowledgable people.

② He delves passionately and blindly into things that interest him. He is always determined to be on the top shelf.

⇒ Once he begins to dig into a project or activity, he becomes blind to everything else to the point of neglecting to feed himself or to tend to his most basic needs. This behaviour is caused by his deep, insidious distrust of others, forcing him to take up all the work by himself. He needs to become more open to and trusting of others and learn to share his workload with friends and co-workers.

③ He becomes agitated and uncontrollably angry when thwarted in his attempts to impose his will on others.

⇒ The Taeyangin is capable of making perceptive and sound judgments most of the times. However, conflicts can arise when he tries to impose his own ideas onto others. With a more receptive and lenient attitude, the situation will improve itself with time.

(2) Diet

① The Taeyangin easily falls prey to escapism and becomes addicted to alcohol.

⇒ Because alcoholic drinks are often served with high calorie side dishes in Korea, alcohol habits can cause drastic weight gain and should be dropped in order to keep in shape. It also deteriorates the mind, and alcoholism is a vice that needs to be fought against through activities that can fend off the addiction, such as music or fine arts.

② Reckless and indiscriminate eating can cause obesity, constipation, leg weakness, or vomiting in Taeyangin.

⇒ The Taeyangin must learn to choose his food carefully. Edible foods include plain-tasting foods, vegetables, seafood, and shellfish. Meat and wheat products are strictly forbidden.

(3) Exercise

① Exercises that generate heavy sweating are not recommended.

⇒ Physical training that involves both the mind and body are most suitable for the Taeyangin, such as qigong, martial arts (wushu), and swordcraft.

② Exercise therapy for the Taeyangin

⇒ The typical "yang" somatotype is characterized by a well developed upper body and a poorly developed lower body. Exercises that train the lower back and abdomen are recommended, such as mountain climbing, walking, cycling, soccer, and basketball.

⇒ The Taeyangin have a well developed back and a poorly

developed abdomen. Stretching and flexibility exercises are beneficial to the muscles of the back and head, while sit-ups and strengthening exercises can fortify the abdominal muscles.

The Soyangin, being of a "hot" constitution, meaning that they have a lost of inner "heat" in the system, are mostly thin, but some are born with a sturdy and stocky somatotype, the type of body frame that often leads to obesity. Their extroverted and energetic personality can contribute in keeping the flab in check, but the typical recklessness and tendency to binge need to be curbed in order to lose weight. Dieting plans should be combined with adequate physical training for the best results.

(1) Lifestyle

① The Soyangin is generally more focused on the outside of matters than the inside.
⇒ His life revolves around the idea that "looks are what counts," an outlook that can cause him to resort to immoderate, extreme dieting that can decay and damage health. He should seek for an expert's advice and keep to a well planned and healthy weight loss program.

② He is rash, impulsive and reckless, always impatient for quick results.
⇒ Excessive pressure and tension will only produce fruitless, rushed, and sloppy results. Reflect and meditate before rushing into any premature actions.

③ He dabbles with numerous activities that he seldom sticks through to the end.
⇒ The Soyangin is one of those people who try every single diet regime in sight without matching success, largely due to lack of the willpower to go through with it.

(2) Diet

① The Soyangin practically sweeps up the food on the table.
⇒ His impatient temperament prompts him to rush through his meals. Measures need to be taken to break this habit; do not eat alone, find partners to talk to during meals, and chew each mouthful at least ten times before swallowing.

② He habitually eats a very large amount in a very short time span.
⇒ Many Soyangin are accustomed to an unhealthy eating habit that can easily lead to obesity: they skip breakfast, opt for a light lunch-snack, and feast on a sumptuous dinner. To correct this, they should eat regularly on meals with limited calorie intake that best suits their lifestyles.

③ The typical irregular and often-skipped meals can cause constipation, heartburn, and headaches.
⇒ In clinical practice, Soyangin who suffer from gastric and duodenal ulcer caused by irregular eating habits are commonplace. The importance of regularity in eating can never be overemphasized. And especially for Soyangin, regular bowel habits, say, one trip to the toilet each morning, are essential not only for losing weight but also for general health.

(3) Exercise

① The typical Soyangin rarely perseveres through a project.
⇒ Forget the grand plans and stick to a steady, easy program. The Soyangin can succeed in losing weight only if he continues to exercise for at least 3 months, 3 times a week.

② Exercise therapy for the Soyangin
⇒ The general exercises beneficial for the "yang" types are recommended, as with the Taeyangin.

⇒ The typical Soyangin frame is characterized by well developed chest and poorly developed lower back, hip, and lower limbs, suggesting stretching and flexibility exercises for the chest and muscle-building work-outs for the lower back, hip, and lower limbs.

The Taeeumin are generally overweight. Some of the contributing factors to this include their characteristic gluttonous greed for food and the obstinate, lazy attributes. They lean toward emotional eating, a tendency that needs to be tackled through a psychological approach. Another suggested solution to the Taeumin weight problem is a more active social life brimming with fun-filled hobbies and leisure activities, possibly combined with a proper physical training.

1) Lifestyle

① The Taeeumin has a covetous nature that clutches at material goods, both his own and others.'

⇒ There are many forms of greed, but the Taeeumin is especially vulnerable to material greed (for food and money), which leads to obesity of the body and condescension of the mind. He must always be checked and reminded of moderation.

② He is typically lazy and idle.

⇒ The Taeeumin needs to learn the virtues of organization and premeditation. Cultivate diligence and beware of laziness.

2) Diet

① The Taeeumin are one of those who claim to be capable of "gaining weight with just a glass of water."

⇒ The "gaining weight with just a glass of water" phase emerges only after the person has arrived at a certain level of obesity.

In other words, the overweight person recognizes pathological patterns only after gaining weight and going through repeated episodes of constipation and menstrual disruption. A specialist should be consulted to check for underlying pathological conditions of the digestive or urogenital system.

② Avoid nighttime snacks.
⇒ The Taeeumin simply cannot resist hunger. They claim that only meat can sate their hunger pangs and that their empty stomach needs to be appeased with a night snack before giving way to sleep. However, contrary to the keen appetite, the Taeeumin is capable of subsisting on two meals a day, and they are advised to go to bed with an empty stomach if they truly desire to lose weight.

③ Avoid overeating.
⇒ Gluttony is a common vice among Taeeumin whose appetites are sated only with a full, sumptuous meal. A firm limit should be set on the amount of food to be ingested for each meal.

3) Exercise

① The Taeeumin has a predilection for the sedentary lifestyle.
⇒ The typical Taeeumin is either sluggish or lazy, more often found immersed in computer games than at the gym. He needs to get involved in energetic physical training sessions to sweat off the extra flab.

② A wider range of physical activities is recommended.
⇒ Being of a rather bovine nature, the Taeeumin rarely introduces change or excitement into his usual routine of exercises without getting some goading from another person. Listen to the advices offered by others and give some air of change by exploring new, different activities that can

spice up the routine work-outs and make use of all the muscles of the body proportionately.

③ Exercise therapy for the Taeeumin

⇒ The Taeeumin exhibits the characteristics of a typical "yin" type body frame, displaying sturdy lower body topped by relatively frail upper body. Activities like swimming, jogging, tennis, badminton, golf, baseball, and volleyball can strengthen the chest, back, and upper limbs.

⇒ The well developed abdomen typical of the Taeeumin would benefit from stretching and flexibility exercises, whereas the poorly developed head and back, also typical of the Taeeuminin, should be fortified through muscle-strengthening exercises.

Born with a weak digestive system, the Soeumin is seldom overweight. However, improper eating habits consisting of irregular meals and calorific snacks are capable, coupled with the typical introverted nature and inclination toward inactivity, of causing obesity in rare cases. Therefore, the first stipulations of a successful diet are regular meals, cutting back on snacks, and regular physical activities.

1) Lifestyle

① The Soeumin is one of those who "sweat the small stuff" and become frustrated and stressed out at the most minor of matters.

⇒ Don't heave up the whole bulk of the workload onto your shoulders only to suffer under it. Share and discuss your problems with friends who can help you sort things out.

② The Soeumin is typically indecisive, hesitant, and lacking in confidence.

⇒ Discard your shyness and strive to mingle in with strangers.

The Soeumin is his own worst enemy unless he learns to express himself clearly.

2) Diet

① The Soeumin easily forgoes a meal or two but never skimps on snacks and sweets.

⇒ The Soeumin who complains of being overweight does not necessarily eat a lot. The main reason for his weight gain is his addiction to snacks and sweets. He must never forget that snacks spell disaster for any diet plan.

② He is a choosy eater with a fastidious appetite.

⇒ Nutritionists agree that balanced diets are vital for a healthy life. As long as there are no underlying gastric conditions, a balanced meal containing all food groups should be eaten in order to maintain health. Exceptional foods that should be discarded from the Soeumin diet are cold beverages and wheat products.

3) Exercise

① Do not bite off more than you can chew.

⇒ Unlike other constitutional types, forced or excessive exercise can exert adverse effects and cause fatigue. Work-outs that produce profuse sweating are also harmful. Lighter but more frequent exercise can produce better results.

② Exercise therapy for the Soeumin

⇒ Exercises beneficial for the "yin" types, are recommended, as with the Taeeumin.

⇒ The well developed lower back, hips, and lower limbs of the typical Soeumin somatotype, can benefit from stretching and flexibility exercises, whereas the poorly developed chest can be fortified through muscle-building exercises.

◑ Weight loss fitness essentials

Most of us have at least once wondered 'what's the best way to exercise to keep fat from jumping onto me?'

The fundamental rule no one can escape from is that everyone needs to exercise in moderate amounts. If the exercises are done for weight loss purposes, they need to be combined with food regulation and behavioral adjustments.

It is true that the Taeeumin constitute the larger part of the obese and overweight population, but none of the four types of constitution are immune from obesity.

In the traditional medical view, the accrual of fat in the body tissues can be interpreted as having a quality of "convergence," and the logical solution to obesity must brandish the quality of "divergence." In traditional medicine, producing perspiration has a quality of "diverging" energy. Thus, exercises that produce light sweating can be useful in battling obesity.

We recommend aerobic exercises like walking, jogging, gymnastics, cycling, and swimming. Cycling and swimming put least stress on the joints. Stretching can be handy to do in a small, restricted office room.

In order to burn up the fat resources in the body, the best way to exercise is 'light and long'; strenuous activities crashed in a short time period only uses up mostly glycogen. The most effective training is done with 50% of the maximum exertion, in 30-60 minute sessions, 3-6 days a week.

Many tend to give up exercising after 3-4 weeks of seemingly unrewarding physical training. However, one should continue working out for at least 2-3 months and preferable longer so that exercising can be merged into everyday routine.

The age factor and past medical history must be taken into consideration so that the intensity, time, and frequency of the training can be adjusted gradually and systematically. Also, the physical work-outs must be understood not as a cure-all to all health problems but as a process that can eliminate the underlying risk factors that can threaten health in the future. Exercises should thus be done with patience and persistence.

7. Stroke

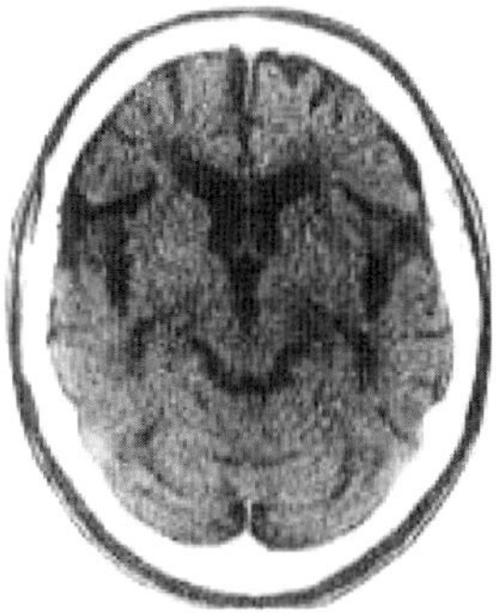

CT image (brain)

Whereas stroke had once been hazily recognized as a group of related signs and symptoms tied together to represent a malady conventionally termed "stroke," it is now diagnosed, through diagnostic tools like CT (computerized tomography) scans or MRI (magnetic resonance imaging), as cerebral hemorrhage or cerebral ischemia. Furthermore, such diagnostic methods can locate the precise site of the anomaly and provide greater accuracy to the diagnosis and prognosis of stroke. However, they are not without loopholes of their own when it comes to capturing the warning signs, preventing the actual attack, and managing the stroke, which have more practical significance than the diagnosis or prognosis.

In Sasang Constitutional Medicine, it is believed that all disorders arise from a disharmony in the Seong (the innermost nature of the minds) and Jeong (the outward emotional disposition), and also that the fundamental cause for the cerebrovascular diseases and other chronic lifestyle diseases lies in the poor management of the underlying risk factors. Furthermore, all the stages of a stroke, including the warning signs, onset, development, treatment, and management, are approached differently for different types of constitution. According to Sasang Constitutional Medicine, a mere improvement of the surface symptoms of a stroke is temporary and volatile in its effect if not accompanied by a matching constitutional improvement.

There are many ways of managing a cerebrovascular disease.

First, a cerebrovascular disease can be controlled through the concept of knowing and doing. In other words, the most basic conditions for a healthy life lies in a proper state of the mind and a proper behaviour of the body befitting the constitutional type. Sasang Constitutional Medicine asserts that diseases are generated by the emotional and mental conflicts within an individual. Thus, efforts must be made to overcome the psychological weaknesses found in each constitutional type in order to prevent diseases beforehand.

Second, stroke can be controlled by modifying one's eating habits. The healthy or unhealthy foods are different for different constitutional types. Supporting evidences were found in a clinical experiment involving Kyung Hee Korean Medical Center patients who were examined for the effects of constitutionally discriminated diet.

Third, Sasang Constitutional Medicine provides a concrete and detailed guideline for managing stroke. One of the important concepts in Sasang Constitutional Medicine is the "Principals of life preservation," or the essential conditions for preserving vitality, and it explains that each type of constitution has different risk factors to watch out for. The same idea is also visible in two other concepts, Constant mindset and State of healthiness. Constant mindset emphasizes the importance of mental stability and peaceful slumber, while the concept of State of healthiness propounds on the attainment of physical well-being through meeting the necessary stipulations of health specific to each of the constitutional types. To be more explicit, easy urination indicates health in the Taeyangin, unconstipated bowel habits indicates health in the Soyangin, unhindered sweating indicates health in the Taeeumin, and smooth digestion indicates health in the Soeumin. Thus, the factors mentioned above (mental stability, sleep, urination, bowel habits, digestion, and sweating) play an important role in the managing of stroke, and therapeutic treatments can be terminated if the six indicators of health are in good condition and under control, even if the patient shows lingering symptoms that seem to point otherwise.

There had been a research on constitutionally discriminated managing of stroke involving 361 stroke patients sent on to the Sasang Department of Kyung Hee Oriental Medical Center throughout the year 1994. The incidence observed in the study was 202 Taeeumin (taking up the largest percentage), 108 Soyangin, and 51 Soeumin. The Taeeumin stroke patients were mostly given treatments to balance their idiosyncratic "dry heat liver pathology." The Taeeumin has a weak circulatory system, a trait that make him particularly prone to chronic lifestyle diseases like cerebrovascular disease, diabetes, and hypertension. A Taeeumin is hence urged to take extra precaution against unnecessary weight gain and constipation (constipation is believed to congest circulation in traditional medicine) and also to avoid excessive tension and stress and the subsequent accumulation of "dry heat." Soyangin stroke patients were given treatments corresponding to their idiosyncratic "fire heat pathology," which typically progresses rapidly but also recovers rapidly, provided he is given the proper treatments. In other words, the Soyangin has a lot of "heat" in the system and thus lacks sufficient "yin material," a condition that demands careful observation. Some of the things the Soyangin can do to prevent stroke is to prevent excessive heat accumulation in the body and to avoid unsettling or distressing the mind. The Soyangin more than any other constitutional types needs to watch out for the signs of heat accumulation in the system, such as constipation, chest constriction, reddening of the face, erythruria, urinary discomfort, or severe thirst. There are comparatively few incidences of the Soeumin stroke. The Soeumin can benefit most by receiving treatments that strengthen the digestive system. In the past, most stroke in Soeumin used to display patterns of congested energy combined with digestive disorders. However, a different pattern has emerged recently amid Soeumin stroke patients, one that is more frequently related to constipation rather than to indigestion.

We have thus discussed the outlines of the different ways of managing stroke for different constitutional types. A cerebrovascular disease is a lifestyle disease that inevitably leaves lingering disabilities and is tagged by risks of recurrence, and the ultimate goal of stroke treatment rests in improving the quality of life for the remaining years through appropriate caretaking rather than in complete recovery. Sasang Constitutional Medicine emphasizes that the risk factors, treatment, and management of stroke has to be differentiated for the different constitutional types.

8. Lifestyle disease

'Lifestyle disease' refers to a group of chronic, persistent, incurable diseases that develop in the later stages of life (around the climacteric), consisting of chronic degenerative diseases that incur either a high death rate or severe disabilities that interfere with normal life.

In order to manage a lifestyle disease, the warning signs must be spotted in the early stages, such as inertia, fatigue, lassitude, insomnia, headache, vertigo, constipation, agitation, uneasiness, and other symptoms indicating a degeneration of health that often first appear in middle-age. More importantly, the person who has spotted the warning signs must respond by making according modifications of lifestyle. Most lifestyle diseases are chronic and not completely curable, necessitating life-long management and rehabilitation. The bottom line in such lifestyle diseases is that prevention is the best and ultimate cure.

There is much in common between the concepts presented in Sasang Constitutional Medicine and the ways of managing lifestyle diseases proposed by Western medicine.

First, Sasang Constitutional Medicine, which emphasizes the importance of preventive care, encourages a wholesome life that follows the regimen of cultivating a healthy mind and harmonizing the Seong (the innermost nature of the minds) and Jeong (the outward emotional disposition). In the same context, modern medicine understands lifestyle disease as being chronic and incurable, qualities that put greater emphasis on preventive care. In light of this similarity, it can be deduced that lifestyle diseases can be controlled by following the guidelines introduced in Sasang Constitutional Medicine.

Second, Sasang Constitutional Medicine assumes that the etiology of all pathological conditions can be found in the context of disharmonies of the Seong and Jeong, a concept closely related to the unique psychosomatic ideology that is expressed succinctly in the phrase, "heal the mind to heal the body." Similarly, mental stress and tension are considered the most responsible factors in causing lifestyle diseases, which is why sufficient rest (in both mental and physical terms) and relaxation are important in preventing lifestyle diseases.

Third, Sasang constitutional medicine proposes different treatment for different constitutional types. This original concept of the "constitutionally discriminated" treatment and management of disease seeks for ways of controlling disease by balancing constitutional weaknesses. Western medicine believes that lifestyle diseases depend largely on genetic and congenital factors. Medical pioneers suggest that promotion of general health and behavioral modification can help combat lifestyle diseases. In light of this, constitutional improvement through constitutionally discriminated treatment can provide an effective way of preventing lifestyle diseases.

As we can see, lifestyle diseases can be prevented by taking precautions and by promoting general health with constitutionally discriminated managing, preventive care, and regimen. Constitutionally discriminated treatment and management can balance the constitutional weaknesses, providing the patients with better ways of managing lifestyle diseases.

9. Spicing up life for householders in their "prime"

Those well versed in their Classics will recall that Confucius was quoted as saying that he had tasted complete tranquility of mind at 40 and came to heavenly realization at 50. However, the modern day householders in their "prime," those between 40 and 60, are perpetually exhausted, both physically and mentally. But these men are the husbands, the fathers, the ones who prop up the society and provide it with security, the ones who should be full of vim and vigour.

Here is what 『Dongeui Bogam』. the greatest classics work in the Korean medical history, has to say about the state of men in their forties and fifties: "...At 40, the internal organs and the twelve meridian systems take a turn from the fullest phase and slowly start to wane. Men hence experience aging of the skin and whitening of the hair and become inactive, finding it easier to be sedentary. At 50, the 'liver energy' dwindles, causing hepatic lobes to thin out and the bile juice to decrease, resulting in failure of sight."

In other words, on a timescale setting 100 as the oldest age possible, Qi and Blood material and the muscles and skeletal structures develop until 30, then the body begins to lose its normal functions, starting from the outside and progressing to the inside, after reaching the apex at 40, then the inside of the body becomes debilitated at 50. Different sources of energy fade out at different ages, the liver energy at 50, the heart energy at 60, the spleen energy at 70, the lung energy at 80, the kidney energy at 90, and finally the Essence and Spirit material disappears at 100, terminating life.

With or without the kind explanations in the book, it is the natural order of life that people are bound to experience mental and physical degradation and slowly degenerate as they age. However, another blazing piece of information impossible to disregard is that the death rate of Koreans in the 40-50 age group is the

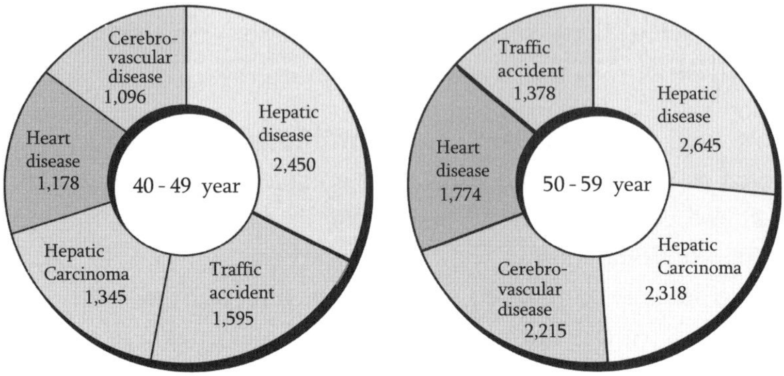

Death rates for 40-59 years old male population (per 100,000)

highest in the world, especially for men. What could be driving them to their deathbeds instead of to the offices or to another place where their wisdom and experience could be doing some good for the society? How can the death rate be lowered?

○ According to recent statistics...

Death rates for the 40-59 year old male population is double that of developed countries and triple the death rate for the female population in the same age group. The main causes of death were found to be hepatic disease, circulatory disease, and traffic accident. Sudden death is responsible for many of the deaths, second only to the traffic accident.

In further analysis, the leading causes of death in the male population in their forties were hepatic disease, traffic accident, and hepatic carcinoma, and for the women in the same age group, cerebrovascular disease, gastric carcinoma, and traffic accident, in that order.

For those in their fifties, the main causes of death appear to be cerebrovascular disease, hepatic disease, and hepatic carcinoma. More specifically, hepatic disease, hepatic carcinoma, and cerebrovascular disease are the leading causes of death for men, and cerebrovascular disease, heart disease, and hepatic disease for women, in that order. Apparently, the difference in age by a decade makes little difference on the causes of death.

Then why are there so many sudden deaths in middle-aged men, presuming that the statistics are accurate? Fundamentally, it all stems from chronic fatigue and stress. They are the ones who built post-war Korea, the ones who sacrificed themselves for the well-being their households and the education of their children, the ones who at the same time are now being attacked from all corners and smothered under the pressure.

However, exerting greater influence than the mental distress are the alcohol and nicotine that they use to escape from it all. The stressful, overloaded working conditions and the abusive drinking and smoking habits all collude together to stack up on the fatigue that in the end crushes them to death. The pathological conditions found in these men can be grouped together under the heading "lifestyle disease," a group of chronic, persistent, and incurable diseases that show its first signs around menopause and andropause.

The so-called "lifestyle disease" pertains to degenerative disease that can progress into geriatric disease if not properly cared for, an inevitable aftermath of physical and mental aging. The best known of these are malignant tumors, hypertension, cardiovascular disease, CVA, diabetes, chronic liver disease, chronic kidney disease, osteoarthritis, etc.

○ The diseases middle-aged men should watch out for are...

First of all, cancer is one of the most common and lethal diseases. Although anti-tumor agents are being discovered one by one, cure for cancer so far still relies heavily on the early detection of the tumor. In other words, the best way to prevent cancer is to have regular medical check-ups and to submit for thorough scrutiny if alerted by suspicious symptoms.

Until recently, gastric carcinoma, hepatic carcinoma, and lung cancer made up the majority of the mainstream cancer in Korea. However, Westernized diet with its typical high-in-fat dishes, environmental pollution, and increase in alcohol consumption have all contributed to the continuous rise in colon cancer, pancreatic cancer, breast cancer, and prostate cancer.

The most common cancer in Korea is the gastric carcinoma. A suggestive reason for this can be found in the typical Korean diet that consists of salted and spiced meat, salted vegetables and fish, and cooked-on-fire meat. More significant that these, however, are the genetic factors. According to statistics, the chances of gastric cancer soars after 50, twice greater for men than for women.

Its first signs are subtle and therefore easily dismissed, often nothing more severe than light indigestion, heartburn, loss of appetite, and other symptoms suggestive of functional dyspepsia or gastritis. However, if such symptoms, however trivial they seem, persist for more than 2-3 weeks, patients are strongly urged to take gastric examination or radiological testing.

Hepatocellular carcinoma and cholangiocarcinoma make up 95% all liver cancers, and hepatocellular carcinoma constitute 90% among the two. The hepatocellular carcinoma is a highly lethal type of cancer, producing the highest death rate among all cancers, and early discovery and treatment are thus important. The incidence of this cancer peaks at 55 and is quadruple that of women in men.

It mostly starts as hepatitis or cirrhosis complication, a reason these disorders of the liver must be screened on regular basis through blood tests and diagnostic imaging.

Recently, a new technique called "fine-needle aspiration biopsy (FNAB)" is sometimes applied, where a small tissue fragment is collected by stabbing the needle into the carcinomatous part of the liver to be examined under the microscope.

The factors that can be held most culpable in causing liver disease is fatigue and alcohol. Our householders in their forties and fifties are perpetually soused in chronic fatigue and incessant drinking. All things considered, it is only natural that hepatic disease places top on the list of causes of death.

One of the ways of preserving a healthy liver is to take sufficient rest and to supply the body with good, nurturing diet containing plentiful of yellow and green vegetables and fruits. It also goes without saying that the middle-aged man must abstain from alcohol and start exercising.

Second, circulatory diseases like stroke or hypertension are another of the more common ailments found in middle-aged men. The factors contributing to the increase of circulatory diseases include greater consumption of animal fat due to westernized diet, obesity, mental stress, increase in smoking, lack of exercise, etc., and with no visible efforts to correct the responsible factors, it can be prospected that the soaring curve will rise even higher.

The stroke, equivalent of the conventional "stroke," occurs when one or more of the blood vessels supplying the brain is blocked or ruptured, thus resulting in partial damage to the brain. The former case is termed "cerebral infarction," and the latter "cerebral hemorrhage."

Hypertension, the underlying pathology in all circulatory diseases including stroke, is the condition of abnormally increased blood pressure in all the arteries of the body. A blood pressure of over 140/90mmHg is generally accepted as hypertension. Hypertension is the main party responsible for cerebrovascular diseases and cardiovascular diseases, and all men between 40 and 60 are urged to keep their blood pressure strictly under control. Some of the things one can do to maintain healthy blood pressure include regular checking of blood pressure, keeping within normal weight range, eating low-salt and low-fat diet, and getting adequate physical training.

Third, the climacteric can be a difficult period, testified by the growing awareness of the "climacteric syndrome." Until recently, climacteric changes were believed to be found exclusively in women in menopause. However, men between 40-60 can also go through climacteric changes that produce mental and physical symptoms as they enter the "andropause."

For women, menopause begins with the termination of menstruation, often marked by hot flashes of heat that rush up to the face, closely followed by cold sweat. These classic signs of the menopause are caused by decreased levels of sex hormone that interfere with normal vascular motility. Other common symptoms of menopause include insomnia, dyspepsia, headache, depression, and osteoporosis.

By contrast, the male climacteric, or andropause, is characterized by severe fatigue and lethargy that are accompanied by loss of libido (sexual drive) and decreased sexual performance (erectile dysfunction). The men experiencing a decrease in testosterone fall into states of dysphoria (restlessness), often becoming agitated and depressed, easily irritated and piqued by small matters. They complain of loss of memory and degraded powers of concentration. Androgen deficiency induces loss of muscle mass and strength that result in skinny limbs, which contrasts acutely with the enlarged belly caused by a redistribution and increase of body fat.

The climacteric symptoms are often treated with hormonal therapy, but more important than the medications is that the man lives an active and fulfilled life, pursuing his fields of interest and keeping in shape through regular exercising.

○ In order to pass the forties and fifties well...

Those in the medical profession agree that regular physical examinations, hobbies or sports that can relieve some of the pressure, abstinence form smoking or drinking, and keeping within normal weight are the things that can lead us to a healthier life.

Along with these, a regular and proper diet is another requirement for a healthy life. Incorrect and baseless beliefs that dietary supplement products or drugs can somehow guarantee health need to be discarded.

A relaxed frame of mind is also important. Most middle-aged men in Korea are horded into business dinners even after work hours instead of being rewarded with the chance to go home and relax. Even at home, many are pressed to be 'the perfect husband and father,' another source of their endless misery. These men need to take a break and spend some quality time just relaxing and regrouping.

Middle aged men can stay healthy only with consistent self-managing, the strong determination to correct the past lifestyle, and love and understanding from his family.

◑ Don't miss out on the medical check-ups

A doctor's appointment is always intimidating, if not scary. Nevertheless, keeping in mind that the forties and fifties is the phase in life with the highest chances of sudden death, a friendlier relationship with your health practitioner may reward you with a healthier old age.

The basic physical exams include height and weight measures, blood pressure (to be measured every year), breast examination (to be examined every year), and a thorough check-up every 2-4 years. Also recommended are thyroid exams (to be done on a yearly basis) and digital rectal examination (DRE, starting at 50). Clinical examinations recommended for middle-aged men are HBV test, cholesterol test (every 5 years), stool examination (every year), AST and ALT (every year), pap smear (every year), gastric endoscopy or fluoroscopy (every 5 years), chest x-ray (every 2 years), breast x-ray (every 2-3 years), etc. Other possible examinations include eye exam, hearing test, various x-rays, pulmonary function test, dental exam, sonography, immunological test, urinalysis, laboratory test, hematological test, etc.

Some may want to consider taking a cancer test. If screen tests show untoward results or if alarming symptoms appear, the patient can submit to a biopsy for an early detection and treatment of the tumor, which is crucial for a successful cancer therapy.

The circulatory system should also be tested for dysfunctions or abnormality. Tests for screening cerebrovascular diseases include CT scan, MRI (or MRA), and cerebroangiography. Cardiovascular diseases can be screened with lab test, electrocardiograph, stress test, 24-hour EKG, and echocardiography. If any of the test results show an abnormality or if same symptoms recur over time, a more thorough examination is required.

10. Treatment of hepatic disease in Korean Medicine

肝臟

『Dongeui Bogam』, the most famous classics work in the history of Korean medicine, draws attention to the parallels found in nature that are related to the "liver system" in our body:

Liver

"There is wind in heaven and trees on earth as there is muscle in the body, liver in the organs, blueness in the hue, eyes in the sensory organs, anger in the desires, sourness in the tastes..."

According to this script, irritable temper, yellow-tinged eyes, and incessant fatigue are all signs of a pathological condition of the hepatic system. In other words, Korean Medicine explains that an irritable temper, the eyes, fatigue, and liver are all interrelated in a single pathological condition, and this association can be applied in diagnosing hepatic disorders. For example, the conditions of the eye, skin tone, appetite, emotional state, digestion, bowel habits, and urinary signs are all indicators of health that reflect the liver functions.

In Korean Medicine, excessive alcohol, overworked states, intemperate eating, angry mood swings, frequent injury, stress, and licentious sex life are rated among the main contributors to liver disease. These factors either directly affect the liver or induce liver disease indirectly by influencing the "spleen system" or the "kidney system" which are related to the "liver system."

○ Treatment of liver disease in traditional medicine

There are two cardinal signs that point to a hepatic disorder. One, hepatic anomaly is outwardly manifested by a bluish countenance, irritable mood, and inexplicable predilection for cleanliness and tidiness. Second, there are inwardly felt symptoms that indicate a liver disease: the patient complains of a throbbing spot left to the navel, heavier-feeling limbs, discomfort in bowel movements and urination, and more frequent injury to the tendons.

Korean Medicine also describes other signs of hepatic disease including headache, vertigo, pain in the flanks, jaundice, abdominal mass, abdominal distension, fatigue, and conditions related to excessive alcohol consumption.

Western diagnostic tools such as the lab test, immunological test, ultrasound, CT, MRI, and biopsy can be used to diagnose viral hepatic disease, toxic hepatitis, fatty liver, alcoholic liver disease, cirrhosis, and hepatic carcinoma. If the aforementioned devises are applied in diagnosis together with the traditional diagnostic tools that assess the pulse and the constitutional type, the hepatic symptoms and functions could be scanned and evaluated to produce far more accurate accounts of the progression and prognosis of the liver disease. Traditional treatment revolves around pharmacotherapy and constitutionally discriminated diet therapy. They are sometimes be combined with acupuncture and qigong therapy.

The crux of treating viral hepatic disease lies in swift suppression of the inflammation, something that can be done in traditional medicine by cooling the "liver heat" and sending it out through urine. The primary and most basic medicaments used for viral hepatic disease are *Injinohryeong-san* (茵蔯五苓散)[2] and *Gwanjung-tang* (寬中湯),[3] and additional or unnecessary herbs can be

---

2 Injinohryeong-san (茵蔯五苓散)
Artemisia capillaris Thunberg (茵蔯蒿) 10g, Alisma canaliculatum A. Braun & C. D. Bouche (澤瀉) 10g, Poria cocos Wolf (= Pachyma hoelen Rumph) (茯苓) 6g, Atractylodes japonica Koidz (白朮) 6g, Polyporus umbellatus Fries (豬苓) 6g, Cinnamomum cassia (Presl.) Blume (肉桂) 2g.

added or taken out according to the symptoms and the constitutional type. Toxic hepatitis can be treated most effectively by eliminating the factors that could have caused the toxicosis, such as recently ingested drugs or allergic food. Having thus prevented further complication, medications that help the bile circulation (enterohepatic circulation of bile acid) or protect the hepatic cells can be administered. Getting proper nourishment and adequate rest is also important.

Fatty liver is treated in traditional medicine by eliminating the "damp heat." If the "damp heat" is caused by excessive ingestion of alcohol, the "dampness" and "heat" should be discharged from the system through sweat and urine. Eating too much rich and thick food can also cause fatty liver, in which case food restriction and weight control measures are required.

If the fatty liver is a complication of a preexisting diabetic condition, treating the diabetes comes foremost in priority. Some of the herbal medicaments that have shown outstanding results in clinical cases are *Galhwahaeseong-tang* (葛花解酲湯),[4] *Daegeumeumja* (對金飮子),[5] *Taeeuminjowi-tang* (Taeeumin) (太陰調胃湯),[6]

---

3 Gwanjung-tang (寬中湯)
Zingiber officinale (L.) Rosc (乾薑) 4g, Alpinia officinarum Hance (良薑) 4g, Citrus unshiu Markovich (靑皮) 4g, Trogopterorum Faeces (五靈脂) 4g, Zingiber nigrum Gaertner (益智仁) 4g.

4 Galhwahaeseong-tang (葛花解酲湯)
Pueraria thunbergiana Benth. (葛花) 20g, Amomum xanthioides Walk. ex Bak. (縮砂) 20g, Amomum cardamomum Linne (白荳蔻) 20g, Citrus unshiu Markovich (靑皮) 12g, Atractylodes japonica Koidz (白朮) 8g, Zingiber officinale (L.) Rosc (乾生薑) 8g, Massa medicata Fermentata (神麴) 8g, Alisma canaliculatum A. Braun & C. D. Bouche (澤瀉) 8g, Panax ginseng C.A. Meyer (人蔘) 6g, Polyporus umbellatus Fries (豬苓) 6g, Poria cocos Wolf (= Pachyma hoelen Rumph) (茯苓) 6g, Citrus unshiu Markovich (橘皮) 6g, Inula helenium L. (木香) 2g.

5 Daegeumeumja (對金飮子)
Citrus unshiu Markovich (陳皮) 12g, Machilus thunbergii Sieb. et Zucc. (厚朴) 3g, Rhizoma Atractylodis (蒼朮) 3g, Glycyrrhiza uralensis Fisher (甘草) 3g, Zingiber officinale (L.) Rosc (生薑) 3g.

6 Taeeuminjowi-tang (Taeeumin) (太陰調胃湯)
Coix lachryma-jobi L. var. ma-yuen Stapf (薏苡仁) 12g, Castanea crenata Sieb. et Zuce (乾栗) 12g, Raphanus sativus L (蘿葍子) 8g, Schizandra chinensis Baillon (五味子) 4g, Ophiopogon chekiangensis K. Kimura et H. Migo (麥門冬) 4g, Acorus gramineus Soland (石菖蒲) 4g, Platycodon gtandiflorum A. DC (桔梗) 4g, Ephedra sinica Stapt

Hyangsayangwi-tang (Soeumin) (香砂養胃湯),[7] and *Moktongdaean-tang* (Soyangin) (木通大安湯).[8]

Alcoholic liver disease caused by excessive drinking should be treated by discharging the waste material through sweat and urine. Some of the herbal medicaments proven to have such effects include *Galhwahaeseong-tang* (葛花解酲湯), *Jujeunghwang-ryeonhwan* (酒蒸黃連丸),[9] *Daegeumeumja* (對金飮子), and Yeol-dahanso-tang (Tae-eumin) (熱多寒少湯),[10] and additional herbs, such as Puerariae Radix (葛根), Poria (赤茯苓), Pinelliae Rhizoma (半夏), Coptidis Rhizoma (黃連), Allii Radix (葱白), Sanguisorbae Radix (地楡), Schizonepetae Herba (荊芥), and Lycii Fructs (枸杞子), can be added depending on the patient's symptoms and constitutional type.

Hepatic carcinoma is difficult to detect in the early stages because of its vague and inconspicuous symptoms. Western medicine treats it with embolization, radiotherapy, surgical incision, chemotherapy, and other techniques that can be employed to suppress the cancer. The results can improve when complemented with traditional herbal medicaments that fortify the immune

---

(麻黃) 4g.

7 Hyangsayangwi-tang (Soeumin) (香砂養胃湯)
Panax ginseng C. A. Meyer (人蔘) 4g, Atractylodes japonica Koidz (白朮) 4g, Paeonia lactiflor Pallas (= P. albiflora Pall) (白芍藥) 4g, Glycyrrhiza uralensis Fisher (炙甘草) 4g, Pinellia ternata Breitenbach (半夏) 4g, Cyperus rotundus L (香附子) 4g, Citrus unshiu Markovich (陳皮) 4g, Zingiber officinale (L.) Rosc (乾薑) 4g, Crataegus pinnatifida Bunge (山楂肉) 4g, Amomum xanthioides Walk. ex Bak. (砂仁) 4g, Amomum cardamomum Linne (白荳蔲) 4g.

8 Moktongdaean-tang (Soyangin) (木通大安湯)
Akebia quinata Decaisne (木通) 24g, Rehmnnia glutinosa Libosch (= R. glutinosa libosch. var. purpurea Makino) (生地黃) 16g, Sophora flavescens Aiton (= S. flavescens Ait. Var.angustifolia Kitagawa) (苦蔘) 8g, Notopterygium forbesii Boiss (羌活) 4g, Coptis japonica (黃蓮) 4g, Plantago asiatic Linn (車前子) 4g.

9 Jujeunghwangryeon-hwan (酒蒸黃連丸)
Coptis japonica (黃蓮) 160g

10 Yeoldahanso-tang (Taeeumin) (熱多寒少湯)
Pueraria lobata Ohwi (= P. hirsuta Matsum., P. thunbergiana Benth.) (葛根) 16g, Scutellaria baicalensis Georgi (黃芩) 8g, Angelica tenuissima Nakai (藁本) 8g, Raphanus sativus L (蘿菖子) 4g, Platycodon gtandiflorum A. DC (桔梗) 4g, Cimicifuga dahurica Maxim (升麻) 4g, Angelica dahurica (Fisch. ex Hoffm) Benth. Et Hook. f (白芷) 4g.

system and improve circulation. Moreover, the herbs discovered to have antitumor effects can prolong life and alleviate some of the sideeffects of the therapy.

○ Good nutrition and sufficient rest are the best remedy

Regardless of which disease, prevention is the best cure. In the same way, the sometimes life-threatening liver disease can be best fended off with a wholesome lifestyle of nutritional diet and sufficient rest. A balanced diet of protein, carbohydrate, minerals and vitamins are most beneficial, which can be improved by throwing in some yellow-green vegetables, fruits, fish, and shellfish.

Fatigue spells red alert for the liver, and no time should be lost to visit a physician if inexplicable feelings of fatigue recur. 20-30% of all fatigue cases are caused by definite physical anomalies, but the rest are largely psychological or caused by an underlying hysteria. For the latter, a light, relaxing bath, or a quick, invigorating hour at the gym, or a massage session to relax the tension in the shoulders, back, and feet can shake away most of the fatigue. Some may find certain postures invigorating, like standing on the head, and certain herbal teas are known to have good anti-fatigue effects.

Those who find themselves waking up in the morning with a severe hangover all too often may prefer arrowroot tea, bean sprout soup, or Artemisiae capillaris Herba. Anemia can also cause fatigue, in which case other kinds of tea can be used, like Angelicae gigantis Radix tea, *Ssanghwa* tea (雙和茶),[11] and *Shibjeondaebo-tang* tea (十全大補湯茶).[12] Cassaiae Semen tea is recom-

---

11 Ssanghwa tea (雙和茶)
Paeonia lactiflor Pallas (= P. albiflora Pall) (白芍藥) 10g, Rehmannia glutinosa Libosch (= R. glutinosa libosch. var. purpurea Makino) (熟地黃) 4g, Astragalus membranaceus Bunge (黃芪) 4g, Angelica acutiloba Kitagawa (當歸) 4g, Cnidium officinale Makino (川芎) 4g, Cinnamomum cassia (Presl.) Blume (桂枝) 3g, Glycyrrhiza uralensis Fisher (甘草) 3g, Zingiber officinale (L.) Rosc (生薑) 3g, Zizyphus jujuba Miller var.inermis Rehder (= Z.jujuba Mill.) (大棗) 3g.

12 Shibjeondaebo-tang tea (十全大補湯茶)
Panax ginseng C. A. Meyer (人參) 4.5g, Atractylodes japonica Koidz (白朮) 4.5g, Poria

mended for patients with easily-tiring eyes and constipation, and tea made from Lycii Fructus or Mume Fructus are beneficial for patients who are afflicted with light but constant cough spells and frequent darkening of the face. Restless sleep, lower back pain, and tinnitus can be allayed with Epimedii Herba or Eucommiae Cortex. Indigestion and retching can be improved with ginseng, ginger, or yuzu tea.

The best ways of maintaining a healthy liver are abstaining from alcohol (giving the liver sufficient time to recover from alcoholic fatty liver), absorbing sufficient and balanced nutrition, and adhering to a strict, constitutionally discriminated diet.

Alcohol can affect the liver directly. Moderate amounts of it can ameliorate circulation and stimulate the cerebrum (which explains the elevated moods produced by drinking), but excessive ingestion can exert noxious effects on the liver and injure it. Traditional medicine has described the below guidelines to be followed when drinking:

① Do not ingest strong liquor with wheat.
② Persons with pale, white countenance are forbidden to drink excessively.
③ Keep within three glasses.
④ Do not gulp or drink in haste.
⑤ If you have imbibed too much alcohol, throw it up immediately.
⑥ Do not travel long distances after drinking.
⑦ Do not have sex after drinking.
⑧ Avoid cold or windy places after drinking.

---

cocos Wolf (= Pachyma hoelen Rumph) (茯笭) 4.5g, Glycyrrhiza uralensis Fisher (甘草) 4.5g, Rehmnnia glutinosa Libosch (=R. glutinosa libosch. var. purpurea Makino) (熟地黃) 4.5g, Paeonia lactiflor Pallas (= P. albiflora Pall) (白芍藥) 4.5g, Cnidium officinale Makino (川芎) 4.5g, Angelica acutiloba Kitagawa (當歸) 4.5g, Astragalus membranaceus Bunge (黃芪) 4g, Cinnamomum cassia (Presl.) Blume (肉桂) 4g, Zizyphus jujuba Miller var. inermis Rehder (= Z.jujuba Mill.) (大棗) 4g, Zingiber officinale (L.) Rosc (生薑) 4g.

11. Diabetes

○ Polyphagia (excessive hunger) and polydypsia (frequent thrist)

消渴

消渴之源

內經曰二陽結謂之消註曰一
大腸俱熱結也腸胃藏熱則喜
手陽明大腸主津液所生病熱則目
液不足也足陽明胃主血所生病熱
血中伏火是血不足也結者津液不
皆燥熱爲病也○消者燒也如火
也○心移寒於肺爲肺消肺消者
不治註曰金受火邪肺藏消爍氣無

Sogal (Diabetes)

Korean Medicine has a special terminology for the diabetes, "Sogal," meaning a state of insatiable hunger (eating unusually large amounts which are digested readily) and unquenched thrist (despite drinking incessantly). It can easily be seen that "Sogal" refers to a larger group of chronic wasting diseases that encompasses the diabetes.

The ancient medical script 『Internal Classics』 records, "The disease commonly called 'So' can be found in overweight members of the nobility who indulge in rich and sweet foods." Similarly, modern society abounds with automatized devices that free us from much of the physical toils, delicious and fattening foods entice us from every corner, and no sector of the structure lacks its fair share of competition and psychological pressure. In other words, no one living in the modern society is entirely safe from the menaces of diabetes.

In Korea, there is an estimated diabetic population of 2 million. The numbers had rocketed in the eighties and now a grand 5% of the whole Korean population are suffering from the malady. More specially, adult diabetes onset soars after 40, and over half of those afflicted are either overweight or obese, implicating the importance of everyday habits and lifestyle for the prevention and managing of diabetes.

According to Korean Medicine, diabetes (or "Sogal") is divided into three different types: "Sangso," "Jungso," and "Haso."

"Sangso," which directly translates as "upper wasting," is characterized by reddening and cracking of the tongue, severe dryness of the mouth, and frequent thirst, all of which can be interpreted in traditional medicine as evidences of "heart heat."

"Jungso," which directly translates as "middle wasting," stems from congested heat in the stomach and the large intestine. Dysfunctions of the large intestine, an organ traditionally viewed as the main regulator of fluid in the body, induces polyuria and dryness of the mouth, evidence of dehydration. Concentrated heat in the stomach, an organ responsible for providing the basic materials for hematopoiesis, causes excessive and insatiable hunger, sever fatigue, and hypobulia (diminished ability to act or decide).

"Haso," which directly translates as "lower wasting," exhibits signs of nephric malfunctioning, such as oily urine, drying of the earlobe, and thinning of legs. The "Sogal" gradually progresses in the order of "Sangso"→"Jungso"→"Haso," culminating over time in a severe condition that accompanies serious complications.

○ Adequate diet and proper physical training are essential

After diabetes is diagnosed through routine lab tests, therapeutic measures, like dietary therapy, physical training, and behavioral adjustment, can be taken depending on the severity of the condition. Diet and nutritional control are the most basic elements in the prevention and treatment of diabetes. Without proper dietary and nutritional monitoring, drugs alone cannot fight the disease effectively.

Diet management for diabetics in Korean Medicine aims at controlling weight, maintaining normal blood lipid levels, retaining normal blood pressure, and preventing further complication. Although many believe that all diabetics are required to cut back on their food intake, dietary restriction is applied only to obese diabetics. In fact, juvenile diabetes and malnutrition-related diabetes mellitus (MRDM) patients are urged to keep themselves well nourished at the same time as keeping blood glucose levels consistent.

Also, diabetics are discouraged against eating only the foods known to lower blood sugar. Indeed, we recommend a balanced diet that attends to the general conditions of the patient. A proper diet would supply the body with just the right amount of calories for the intended weight level, limiting monosaccharides to 10% of the total carbohydrate, and cut back on fat, saturated fat, cholesterol, and salt to the minimum, instead maximizing the fiber content. We also recommend constitutionally discriminated diet managing for diabetics.

For the successful maintenance of proper diet management, the dietitian or whoever preparing the meals must take into consideration the individual particularities and the preferred choice of food. The diabetic hence needs to communicate with the dietitian or physician about his eating habits and personal life in order to be diagnosed with the most compatible diet.

Equally important as the diet in the managing of diabetes is exercising. The type, intensity, duration, and frequency of the exercise need to be determined with the help of a specialist that best attends to the constitutional needs and the severity of the diabetic conditions. The usual variety of physical training offered for the diabetic is fine, but the more adventurous ones may wish to try qigong for breath control, mind-building, and physical health. Approximately 200-300kcal should be burned a day through exercising, depending on the individual health conditions and strength. Physical training combined with diet management will produce better results.

An ideal training scheme would contain 80% aerobic exercise, 15% strength training, and 5% flexibility exercise. The program should not be strenuous, just enough to make the person a little short of breath or to sweat in the back. Beginners may want to start out lightly, aiming for 40% of maximum exertion.

There is a wide range of possible physical activities for the diabetic that includes walking, power walking, jogging, free gymnastics, cycling, mountain hiking, and swimming. Choose from the whole gamut so that working out can be an enjoyable activity. The exercises should be done at least 3 times, preferably 5 times a week to make a difference, in 30 min sessions, starting 30 min after a meal.

○ Live abstemiously

There are three things in everyday life that the diabetic must avoid. One, beware of alcohol. Second, tone down on your sex life. Third, avoid spiced and salted foods or wheat products. Needless to say, stressful conditions can also threaten the diabetic's general well-being.

Korean Medicine diabetes management consists of the basic dietary control, exercising, and behavioral adjustment, which are combined with acupunctural therapy and herbal medication. The details of the acupunctural therapy and herbal medication depend on the symptoms, constitutional type, and the trifurcated division of "Sogal." For example, "Sangso" is treated by cooling the "heart heat," while "Haso" is treated by extinguishing the heat at the same time as protecting the kidney system.

Individual herbs can be prescribed separately: Ginseng Radix (人蔘), Schizandrae Fructus (五味子), Liriopis Tuber (麥門冬), Asparagi Radix (天門冬), Trichosanthis Radix (天花粉), milk, fresh juice of Rehmanniae Radix (生地黃), and fresh juice of Nelumbinis Rhizomatis Nodus (藕節) are frequently prescribed for "Sangso (上消)"; Gypsum Fibrosum (石膏) and Talcum (滑石) are usually used for "Jungso (中消)"; Phellodendri Cortex (黃柏), Anemarrhenae Rhizoma (知母), Rehmanniae Radix Preparat (熟地黃), and Schizandrae Fructus (五味子) are adopted for "Haso (下消)."

Many diabetics turn to folk medicine or food remedies in hopes of discovering a better way of coping with their condition. Some of the most popular diabetic foods in Korea are silkworm powder, mulberry leaf, common pea, guava, chinese lizard's tail plant (*Saururus chinensis*) wine, japanese angelica tree shoot (fatsia), roasted peanut, pig pancreas, eleuthero (*Acanthopanax senticosus*), chinese fine scaled weather loach, pumpkin, loquat leaf, and kiwi. And indeed, some of the foods, though a tiny fraction in the whole gamut of foods, actually do have hypoglycemic (lowering blood glucose) effects. Still, an unprocessed food can produce unexpected adverse effects, and the foods need to be corroborated by prolonged scientific research to extract the active agents and process them for easy consumption in adequate doses.

Taking the silkworm powder for example, studies have actually shown its hypoglycemic effects, and yet it is deemed by scientists to be inappropriate for actual application for many reasons. It has limitations in effectiveness that are not reflected in its expensive prices, the unrefined powder samples can be adulterated with toxic substances, and patients may make the unwise decision to rely more on the food remedies than on professional medical advice. Prudent decision-making is required.

See Appendix 4 for further information on constitutionally discriminated diet management for diabetics.

12. Integrative medical check-up: Combining the wisdom of the West and East

There are situations when a patient reports of a physical symptom that fails to be confirmed with conventional medical tests. Chances are, he is bound to stray the whole gamut that starts with alternative medicine and ends with folk remedy. By the time his conditions are finally diagnosed in standard medicine, it is already too late to reverse them.

Given the abundance of such cases, patients are urged to take further steps if the routine medical examination fails to reveal the underlying pathology. The integrative medical check-up, which combines the better parts of Western medicine and traditional medicine, is proposed as a alternative to the conventional physical check-up. This integrative medical check-up can provide an upgraded "tailored check-up" services that can account for the individual idiosyncrasies as well.

○ What is the integrative medical check-up?

In simple words, it is the conventional routine medical check-up topped with traditional medical examination. It encompasses a comprehensive range of examination including those related to the usual routine exams, gynecological tests, GI tract exams, stroke-related tests, and exams for the underaged. Western diagnostic tools like the hematological test, x-ray, and sonography are combined with traditional medical examinations to determine the individual constitutional type and the functions, symptoms, and states of the internal organs.

The Western exams detect physical and structural defects, while the Korean Medicine exams attempt a more holistic approach that reflects the individual physical or constitutional idiosyncrasies in its diagnosis and treatment.

○ What kind of tests are included in the Traditional Korean Medical exam?

The Traditional Medicine part of the check-up is largely divided into three fields: constitutional exam, organ functions exam, and transcranial ultrasound (TCD) exam.

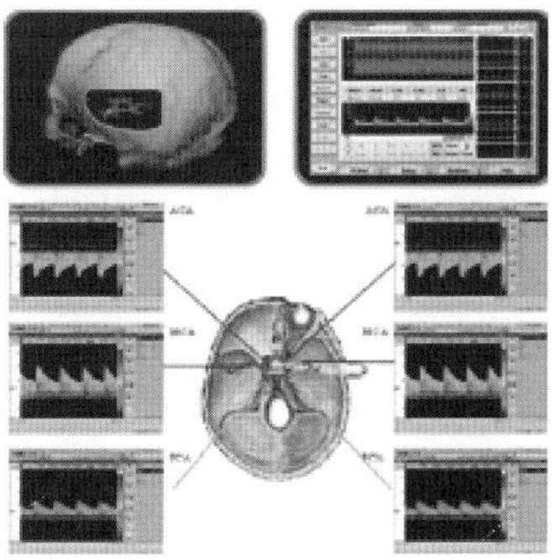

Transcranial Doppler Ultrasound (TCD)

The constitutional exam is rooted in the notion that the personal constitutional type gives hints about the health conditions of the individual. It scans for the diseases most likely to occur in the constitutional type and dysfunctions related to the constitutionally weakest organ. For example, a typical

Taeeumin would have hyperactive hepatic functions and conversely weak respiratory system, making him susceptible to respiratory diseases and prone to obesity.

The individual constitutional type is diagnosed through questionnaire sheets, morphological assessment (the morphological shape, measurements of the facial features and the somatotype), and evaluation of the standing posture.

The organ functions exam is operated in order to determine the functions and conditions of each organ, using various tests like Yangdorak, pulse testing, and other biofunctional tests.

Thermographs using digital infrared thermal imaging (DITI) can be especially useful. It can efficiently capture abnormal heat or coldness and painful spots in lower back or other parts of the body that are difficult to spot with X-ray or CT.

The pulse examinations included in the organ functions exam resembles the traditional "pulse diagnosis," a form of palpation in which the traditional physician presses on a place on the wrist to feel the pulse and perceive the pathological conditions of the body. The pulse analyzing devices now used in medical centers convert the pulse waves into electric waves and store it as a part of a huge database, thus allowing for objective evaluation of diseases in patients.

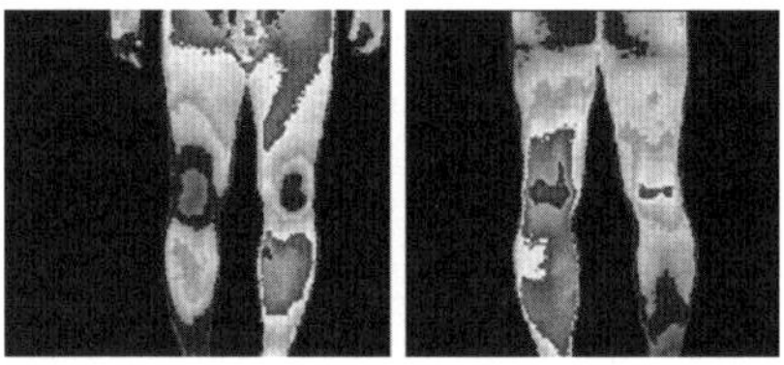

Digital Infrared Thermal Imaging (DITI)

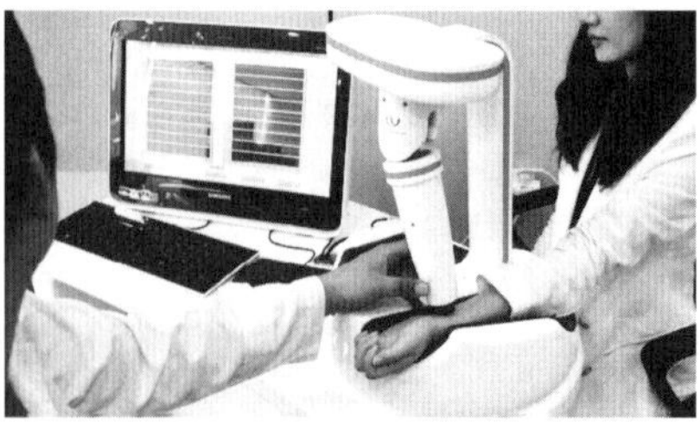

Pulse testing

Lastly, transcranial Doppler ultrasound (TCD) can be used to survey the cerebral bloodstream to examine the risks of stroke and other cerebral diseases.

○ Advantages of having a Korean Medical exam

To be clear, a traditional exam by no means captures diseases more expediently than a Western one. However, the integrative medical check-up can combine the better parts of the two exams to produce more accurate results.

Korean Medicine provides information on the possible preventive measures for specific diseases and the likelihood of becoming inflicted with a certain disease by determining the constitutional type of an individual. Moreover, its holistic approach to health provides information on proper eating habits and healthy lifestyle and the risk factors involved, allowing us to prevent and prepare for future diseases.

Archives: Alcohol and constitutional type

The following are excerpts from a slot on "Ask Us Anything" that aired on KBS (Korean Broadcasting System), February 21, 2002 (10:00-10:50). The guests appearing on the show are Lee Eui-ju and Prof. Kim.

1. Alcohol and constitutional type

Host 2: Well, this piece of information declares that an average Korean adult drinks 1 bottle of shipped liquor, 8 bottles of soju, and 103 bottles of beer a year. Wow. Are there people who just sit around drinking all day or what? You know, this make me think Koreans are born with the no-booze-can-hammer-me genes. Were we really made to take all this alcohol?

Lee: Actually, the opposite is true. Most Koreans cannot stand alcohol that well. Scientifically speaking, being able or unable to stand alcohol depends on the genes that code the enzymes that participate in alcohol detoxification, known as aldehyde dehydrogenase (ALDH). And the Korean people have low levels of ALDH activity. ALDH levels can differ among peoples and persons depending on the genetic makeup. People in Europe are known to have greater levels of ALDH than Koreans, Chinese, or Japanese. A low ALDH level means that the alcohol ingested more or less stacks up in the system in the unfiltered form of aldehyde.

Panelist: And yet, the whole nation's drinking bottoms up, at least according to the statistics. I guess this goes to show we all need a little bit of alcohol to help us through life... Well, in contrast some people purportedly claim that striding across a barley field is enough to make them feel dizzy, would you say the constitutional makeup has something to do with how much alcohol you can tolerate?

Kim: Well, I would say the issue is less about whether someone is compatible with alcohol or not than about alcohol being more harmful or less harmful to someone. Alcohol can affect us in two parts: the mind and the body. Some people receive greatest damage on the nervous system, others on the hepatic and digestive systems. Of course, in the end, the mind and body deteriorate together. If your alcohol tolerance has gradually increased via a drinking habit, it only means that your body loosened up and let its guard down to the frequented alcohol. Loosening up means that your defense mechanism is giving up on labeling alcohol as a noxious substance. And yes, the constitutional element does play a role your alcohol tolerance, and the constitutional makeup can be discriminated by careful observation of how a person reacts to drinking.

2. Alcohol and gender difference

Host 1: How does alcohol act on men and women? Increase in the female drinking population is really becoming an issue nowadays.

Lee: The enzyme that acts in the first stages of alcohol decomposition, the alcohol dehydrogenase, also exists on the gastric walls. In other words, alcohol is partly broken down in the stomach. However, women have less alcohol dehydrogenase than men. Because of this difference, whereas a portion of the ingested alcohol is initially decomposed inside the stomach before absorption in men, in women the whole unfiltered bulk is absorbed directly. This provides an explanation as to why women get drunk faster than men.

Host 2: So, you're saying drinking is far more harmful to women than men?

Lee: Yes, because women have less alcohol dehydrogenase and more fat tissue, which takes them longer to eliminate alcohol from their system. A recent foreign study has discovered that it takes 15 years of 5 glasses of alcoholic beverage every day to induce cirrhosis in men, whereas it takes only 10 years of 3 glasses every day in women.

3. People who redden with drinking

Panelist: You know, I've seen people who turn bright red after a glass and others who simply pale out. They say they are just born that way. Are there any scientific explanations for this?

Kim: If there is a problem with the absorption that causes the alcohol to become congested in the stomach, the face becomes pale the way it does when food is congested. If the detoxifying capacity of the liver is reduced, the blood alcohol levels remain high for a longer period of time, which causes the face to blush.

4. Beer or soju? Let your constitutional type decide

Host 1: So, how could you know what kind of alcohol is right for you? I guess there are a few things you could do offhand, like just trying out the stuff or watching for the warning signs of the red face, but seriously, isn't there a simpler way to find out your constitutionally-oriented alcohol preferences?

Kim: In traditional medicine, alcoholic drink is understood as water that contains great heat. Compared with tobacco, which contains "dry heat," alcoholic drink contains "damp heat." Following the logic, drinking is less harmful to those who are predicted to have a "drier constitution," or if drunk on drier, wintry days. People with a "damp constitution" are generally overweight with a lot of flab packed on, according to traditional medicine. These people have greater chances of developing hepatic diseases even if they personally believe that they take alcohol well. The ones who are really "fit" for drinking have a "cold and dry constitution," and these people generally have a lean body frame, according to traditional medicine. But even these people qualify only if their digestive system is healthy and strong. In other words, thin people with a strong GI tract whose constitution is not "hot" are the only ones permitted to drink, at least in the context of natural selection. Nevertheless, a drinking habit is bound to generate a lot of heat in the body system, which can give the person an irritable temperament over time. As an easy example, you'll discover that some of the real "born" drinkers are skinny, gaunt girls who can eat for two and still maintain an excellent figure. They usually find it difficult to drop their drinking habits even after the calories finally catch up, presenting them with a sudden invasion of fat and a "hot constitution." They don't lose their hot temper either.

5. "Healthy" way to drink for different types of constitution

Panelist: I've heard drinking in moderation can actually benefit health if you follow the "right" ways of drinking for your constitutional type. Is that right? Could you explain that for us?

Host 2: Yes, I'd like to hear a detailed explanation, like what kind of drink is good for the Soyangin and Taeyangin, and so on.

Lee: Well, this is how Sasang Constitutional Medicine looks at it. The constitutional type least injured by alcohol is Taeeumin, while the type to take it the hardest are Taeyangin and Soyangin. Of course, any alcohol is off limits irrelevant of the constitutional type if there is an underlying pathological condition that can be affected by it. Taeyangin are compatible with clear strained rice wine, Soyangin with rice wine, Taeeumin with soju, and Soeumin with unrefined rice wine. Traditional medicine employs unique methods to detoxify alcohol—the alcohol waste material is discharged from our body through facilitated urination or sweating, the former applied generally for the "yang" constitution and the latter for the "yin" constitution.

6. Alcohol induced diseases

Host 1: So, in fact there is a "good" way of drinking, and I think the tips for the different constitutional types are great. But surely drinking is closer to a vice than a virtue in the end. I've heard there's a special term in traditional medicine for the group of diseases caused by drinking excessive amounts of alcohol, how's that?

Kim: Yes, I'll explain about that. "Disease caused by drinking excessive amounts of alcohol" refers to the dysfunctional conditions of the liver, stomach, and large intestine that is caused by excessive alcohol. Some of the modern day diseases included in "Disease caused by drinking excessive amounts of alcohol" are alcohol induced liver disorders like alcoholic fatty liver and alcoholic hepatitis or exacerbated GI tract disorders like gastritis, gastric ulcer, gastroptosis, and bowel malfunctioning.

Panelist: (taking the VCR) I'm guessing that hepatic diseases are one of the worst conditions that can result from drinking. Would there be some sort of a checklist for the warning signs that can alert us?

Kim: Sure, I've got some of them listed here:

① Spider angiomas (spider-shaped red mark) on the chest and back.
② Reddening in the palm sides of the thumb or the small finger (palmar erythema).
③ Abnormal yellow discoloration of the skin or the whites of the eyes.
④ Spoon nails (koilonychia) with raised edges. Whitening or cracking of the nail can also occur.
⑤ Lower extremity swelling.
⑥ Sudden intolerance to alcohol.
⑦ Pain or swelling in the right upper abdomen and flanks.
⑧ Firm swelling felt upon pressing below the right or left rib cage.
⑨ Brownish spots and blemishes on the skin unrelated to tanning.
⑩ Decreased appetite.

My patients would come to me with some of these symptoms and signs and ask me what the problem is. Almost invariably, it's alcohol, alcohol, alcohol. Alcohol plays such a huge role in so many cases that I usually consider alcohol before anything else.

Host 1: Maybe this has something to do the genetic makeup of the Korean people, because I've heard liver disease is common in Korea, demographically speaking.

Lee: Statistics say that hepatic disease accounts for 22.9 deaths per 100,000, coming fifth in the causes of death, following cancer (122.1 deaths), cerebrovascular disease (73.2 deaths), heart disease (38.5 deaths), and traffic accident (25.4 deaths). What's more serious is that it is the second or third largest cause of death for the middle-aged in their forties and fifties, the people who should be up and running and not dying. HBV patients make up 70% of chronic liver disease patients in Korea, and HCV accounts for 15-20%. Chances of cancerous developments in the hepatitis and cirrhosis patients are 100 times that of a normal person. We encourage people to take hematological tests and abdominal sonography every 3-6 months. Male chronic liver disease patients over 40 are in the highest risk group for hepatic carcinoma.

7. Drinking and allergies

Host 2: Some people say that drinking can develop allergic tendencies, is that true?

Kim: Yes. We've had studies showing us that drinking does in fact make you more allergic. According to one research, people who drink more than one glass a day have heightened levels IgE antibody, a substance that triggers immunological reactions. The higher the IgE level, the more vulnerable the person is to allergic rhinitis, asthma, and eczema. Habitual drinkers are known to have higher levels of IgE.

8. The more you drink, the more you *can* drink

Host 1: They say that the more you drink, the more you *can* drink, meaning that after a period of habitual drinking, you have to imbibe more to get the same effects. What would you say to that?

Lee: This is a phenomenon referred to in the medical circles as "alcohol tolerance." The mechanism behind it is, the more alcohol you ingest, the faster the liver breaks it down, and your central nervous system eventually adapts to the constantly high blood alcohol concentration. However, adapting to high alcohol levels also means that you become dysfunctional in lower alcohol levels. This explains why alcoholics experience withdrawal symptoms when they stop drinking or significantly cut back on their alcohol intake.

9. The must-knows of healthy drinking

Host 2: Drinking is a habit better lost than not, but equally, not drinking is something impossible to do in Korea. Couldn't there be a healthy way of drinking, oxymoronic as it sounds?

Kim: Yes, I think the crux of the problem is, drinking is much more than just a personal problem. It involves social codes as well. It goes without saying that the drinking culture has to go, but we're also saying that drinking "good alcohol" is just as important. By "good drinking," we're talking about avoiding poor quality alcohol that contains impurities, we're talking about choosing brewed or fermented drinks over distilled liquor. A healthy man can disintegrate 160-180g of alcohol (equalling 2 bottles of soju) a day. The maximum quantity of alcohol that can be imbibed a day without injuring the liver is 80g (equal to 1 bottle of soju). The figures are halved for a woman. Keep at least 2-3 days in between your drinks, and don't forget to eat other foods with your drink to protect the stomach walls and the liver. And there are other little things you can do to ease the effects of alcohol: keep your cigarettes firmly in pocket when you're drinking, make sure to leave ample space between you and the bottle as often as humanly feasible, stop it if you find yourself drinking in stressful situations and instead vent it out with a friendly chat with friends or mindless singing, show openly your support and admiration for the non-alcoholic culture with words and matching abstinence, practice swaying movements to put into action when your friends press you with another unwelcome glass.

10. How different constitutional types can escape from hangovers

Panelist: (taking the VCR) Right, no one should be pressed into drinking. Are there any good ways of getting over hangovers?

Lee: I could give you some general tips:

① If there's still some remaining alcohol in the stomach, just throw it up.

② If you can't do that very well, drink some strong, hot honey water or tea.

③ Take a lukewarm bath.

(see explanations on How to Win over Alcohol for Different Constitutional Types)

11. Drinking alcohol to pass over hangovers

Host 2: You know, a whole crowd out there actually say that drinking a little helps them go through hangovers. Seriously, this can't be good, can it?

Kim: You can think of it as a narcotic. The drug abuser can get temporary alleviation of the painful symptoms experienced during withdrawal by taking more drug. This short-lived respite only leads to further complications. It means it's deranging your sensory system enough to inure it to pain. This kind of reckless drinking behavior can bring on other alcohol-related complications, like the Mallory-Weiss syndrome, usually caused by forceful or prolonged vomiting ensuing heavy drinking.

Chapter 5

Health guidelines for children of different constitutional types

1. Raising children the ideal way compatible with their constitutional types

○ Living the way you are made is the first step to health

Korean Medicine emphasizes the value of a well-guided, healthy lifestyle, and it looks for pathogenic factors and therapeutic clues in the person's surroundings and living conditions. This is perhaps why people seek for help in Korean Medicine when conventional Western medicine fails to cure their diseases, or when a disease recurs habitually and inexplicably, or when the pathological conditions fail to be diagnosed but are nevertheless distressing for the patient.

As you see, Korean Medicine places great emphasis on the prevention of diseases, and if this attitude can be adopted in everyday life, the general health would be significantly improved for many. Each individual is born with a unique constitutional makeup, and to every constitutional type there are compatible and incompatible environments. Following the logic, diseases can be prevented by avoiding the injurious environments and the foods incompatible with the individual system.

Then how exactly does this constitutional typology work? More importantly, how can constitutional typology be related with raising children? This years marks the 100th anniversary since Lee Je-ma's demise, perhaps the perfect time to start pondering about how Sasang Constitutional Medicine can be applied in day-to-day life and the general culture.

○ Constitutional type and Sasang Constitutional Medicine

Sasang Constitutional Medicine places individuals in one of the four categories according to the internal makeup that involves four organ systems, each revolving around the lung, the liver, the kidney, and the spleen. There is a subtle balancing of energy between the four systems in each individual that must be maintained for the individual's health and stability. This constitutional makeup is manifested on the exterior in outward characteristics exclusive to that constitutional type (physical attributes, appearances, temperament, personality, etc.). The constitutional differences also determine the most vulnerable organs and pathological propensities unique to that constitutional type.

Determining the constitutional type of a person is no easy feat, having versatile standards that appear erratic to the layman, and only experts specializing in Sasang Constitutional Medicine are capable of accurately deciding the constitutional makeup of an individual.

For adults, constitutional diagnosis begins with the observation of facial features, somatotype, lifestyle, and pathological conditions. The related criteria are described in the table below. But for children in developmental stages, a wholly different set of diagnostic standards exists, partly because the physical structures are not yet fully shaped, and also because their physiology and pathology are poles asunder from that of an adult.

Characteristics of different constitutional types

Constitutional type	Taeyangin	Soyangin	Taeeumin	Soeumin
Facial features	Clear-cut features Penetrating eyes	Razor-sharp chin Thin lips Sharp and strong features	Large, heavyset features Gentle, dignified, and reliable impression	Delicate and well-shaped features
Body frame	Thin, wispy abdomen and legs Well developed neck and shoulders Stalwart physique	Well developed chest Flat in the hips Exceptionally skinny legs	Large frame set Well developed lower back and abdomen Poorly developed around the neck	Well developed hips Frail chest and shoulders
Personality	Scintillating mentation Adventurous Self-righteous Strong masculinity	Cheerful and quick-witted Quick to judge An extrovert	Persistent, tenacious, and active Obstinate and intractable Conservative	Meticulous and always prepared Sensitive An introvert Strong femininity
Common maladies	GERD (Gastroesophageal reflux disease), vomiting, lower limb weakness, lower back pain, muscle disease	Urogenital disorder, lower back pain, insomnia, constipation, neurological disorder, stress-related psychopathy	Respiratory disorder, circulatory disorder, enteropathy, obesity, skin disease	Gastroenteropathy, cold intolerance, allergy, diarrhea
Signs of health	Easy urination	Easy defecation	Profuse sweating	Comfortable digestion

○ Offsprings take after parents

If you insist on knowing your child's constitutional type, look into his/her face. The child is likely to be of the same constitutional type as the parent he/she resembles most in appearance. You can easily guess the child's constitutional type by his/her appearance, somatotype, habits, and common ailments. After all, constitutional makeup, like most other personal traits, is inherited from parents.

This method of working out the offspring's constitutional type shows a 70% accuracy, and it is employed in clinical practice as well by specialists who want extra information for more accurate diagnosis. The remaining 30% depends largely on individual attributes like personality, temperament, somatotype, and other variables, and should be left entirely in the hands of physicians who specialize in Sasang Medicine.

○ Let them be what they were born to be

A person's constitutional type is congenital and cannot be completely altered, the way your blood type cannot be changed. What you can do to lead a healthy life is to live according to your constitutional makeup. Your weaknesses can be balanced and the surrounding environments can be readjusted to make your life more compatible with your original propensities.

Sasang Constitutional Medicine describes how specific foods can either harm or help a constitutional type and insists on imposing food restrictions for each type of constitution. However, a growing child should not be deprived of a full diet that spreads over all the food groups. The only exceptions to this rule are cases when a single food product is found culpable for a recurrent and persist-

Heredity

Family tree

173 clinical data couples

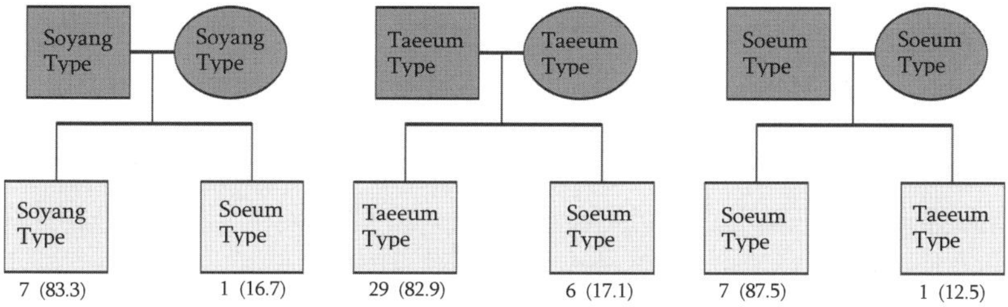

The Sasang Constitution can be believed to be inherited, because the constitutions of children, whose parents have the same constitutions, are mostly the same as their parents.

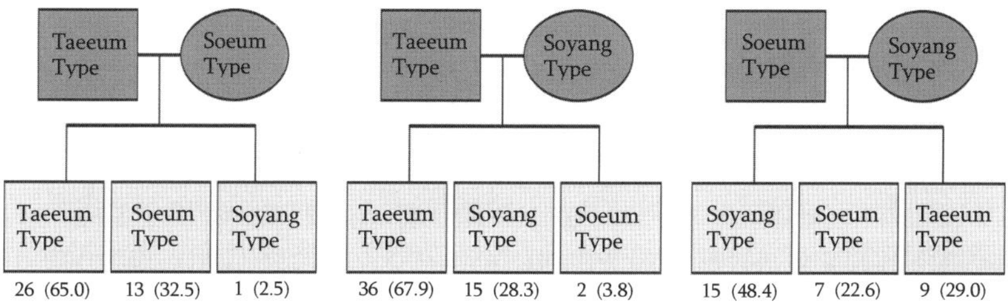

It seems that one constitution is predominant to the other according to combinations of constitutions of constitutions because constitution ratio of children are uniformly 2 to 1 in case the constitutions of parents are not same.

Lee SH, Yoon YS, Kim HG, Kim JY. Clinical Study on the Distribution of Sasang Constitutional between Parents and their offsprings. Korean J of Oriental Physiology and Pathology. 2004 ; 18(6) ; 1904-7.

ent disease, upon which the offending food should be eliminated entirely from the diet. A specialist must be consulted before any final decisions are made.

1) The Taeyangin child

(1) Habits

To cut out their typical headstrong qualities early in childhood, the parents must talk things out with the child reasonably. The child should be offered coherent explanations. Persuasion should be taught, not coercion. The Taeyangin child may listen better to an authority figure (eg. schoolteacher) than to parents, and excessive interference into the kid's life is unhealthy. Playing in groups can cultivate his / her cooperative skills. The Taeyangin child can become intensely immersed in the activities he / she enjoys, enough to even forgo meals, and gentle intervention is required of the parents. The kid will not let go of anything he/ she does not understand, for which he/she will look up to the parents for careful explanation. This critical stage in childhood development should be used to instill an open mind and to teach respect for others.

(2) Eating

A choosy eater, the Taeyangin child often refuses food and has a tendency to vomit. A strict constitutionally discriminated diet is needed to prevent obesity, constipation, lower limb weakness, and vomiting. Meat and wheat products are especially forbidden.

(3) Beneficial foods

Plain-tasting food, vegetables, seafoods, shellfish, octopus, sea mussel, banana, cherry, orange, kiwi, apple, grape, chinese quince, buckwheat, etc.

(4) Exercising

Their upper body is more developed than their lower body, and physical training that aims to strengthen the lower body are most beneficial, such as mountain climbing, walking, cycling, soccer, and basketball, to name a few. The well developed back is outbalanced by the poorly developed abdomen which needs to be toned up with sit-ups and strengthening exercises. Exertive training that triggers off excessive sweating is discouraged. We recommend exercises like qigong, wushu, and swordcraft that can regulate the mind and body at the same time.

2) The Soyangin child

(1) Habits

The Soyangin child is easily riveted by the glimmer and glitz of life, a propensity that can lead to extravagance later in life, requiring parent supervision in shopping. He/she is by nature quick-tempered and needs to be constantly reminded to think before acting. He/she reaps relatively little from all the activities originally intended to accomplished, and it is the role of the parents to teach organization and premeditation by guiding the child and checking for progress through every step.

(2) Eating

A Soyangin child tends to eat too fast in scrunched meals. This bad eating habit needs to be replaced with meals eaten together with family members or friends with plenty of conversation thrown in between. The child must be trained to chew each mouthful at least 10 times before swallowing. A Soyangin child is often afflicted with gastrointestinal diseases which can be prevented in part by adhering to regular meals and cultivating regular bowel habits.

(3) Beneficial foods

Pork, strawberry, watermelon, abalone, sea cucumber, oyster, globefish, red bean, chinese cabbage, cucumber, mung bean, sesame oil, lettuce, etc.

(4) Exercising

The Soyangin is hardly the patient type, and easily attainable goals will help the child keep exercising. The typical Soyangin body frame is characterized by poorly developed lower back, hips, and lower limbs that demand strengthening work-outs, and well developed chest, suggesting stretching and flexibility exercise for this part of the body.

3) The Taeeumin child

(1) Habits

The greed for both food and money typical of a Taeeumin child easily leads to obesity, and self-control must be taught from the earlier stages. Bearing in mind the Taeeumin's sedentary nature and tendency to procrastinate, the parents need to teach their children to plan ahead and cultivate assiduity.

(2) Eating

The Taeeumin child is exceedingly prone to obesity and often suffers from repeated bouts of constipation, indigestion, and fatigue. Eating must therefore be closely monitored, and suppers must be kept light and skimpy. They have to eat huge amounts before feeling satiety, and parents should establish a set limit for the amount of food to be eaten each meal.

(3) Beneficial foods

Beef, adlay, indian millet, sweet potato, walnut, chestnut, wheat, radish, bean, tofu (soybean curd), balloonflower, plum, milk, ginkgo nut, etc.

(4) Exercising

Given their sedentary nature, the Taeeumin children should be urged, and forced if necessary, to participate in active, sweat-generating activities. The child should be given a wide range of activities to choose from so that he/she will not find it tedious. The Taeeumin child is typically more developed in the lower body, which means he/she will benefit most from physical activities like swimming, jogging, tennis, badminton, golf, baseball, and volleyball.

4) The Soeumin child

(1) Habits

The Soeumin child sweats the small stuff. He/she needs to learn to communicate his/her inward feelings more openly and to socialize with more people. Learning to speak out personal likes and dislikes can help the child relax.

(2) Eating

Their diet relies more heavily on the in-between-meal snacks rather than on the actual meals. Plus, Soeumin children are also known for their fastidious and choosy appetite. Parents are required to provide the child with a balanced diet. The only foods to be excluded are cold beverages and wheat products, which are unhealthy for the Soeumin child.

(3) Beneficial foods

Chicken, southern conger eel, glutinous rice, yellow corvina, anchovy, dried alaskan pollack, japanese icefish, freshwater mandarin fish, saury (mackerel pike), laminaria, mandarine, chinese chive, honey, etc.

(4) Exercising

Excessive physical activity is not necessarily healthy for everyone; indeed, it can stack up fatigue for the Soeumin. The more exertive work-outs should be left to others, and the Soeumin child will benefit amply with the lighter, less strenuous programs. The Soeumin body frame is well developed in the hips and less developed in the chest, suggesting stretching and flexibility exercises for the lower back, hips, and legs and muscle-building exercises for the chest.

○ How to recognize your child's illness beforehand

Much of the information on a child's health conditions can be picked up by observing the urine and stool.

Korean Medicine proposes that the respiratory disorders and infectious diseases found commonly in children can usually be detected by observing the urine, while the digestive disorders and internal diseases can mostly be observed in stool conditions. For example, significantly more frequent or less frequent urination indicates possible infection.

Needless to say, factors other than urine frequency must be taken into consideration, like changes in urine volume, color, or smell. For a comprehensive diagnosis, therefore, parents are required above all to observe the usual urine and stool conditions of the child in a healthy state.

2. A weakly child

All parents wish that their child is born healthy with a naturally strong immune system that ensures immunity against diseases. Among the many responsibilities entitled to the parent in raising a child, the importance of feeding and providing the right kind of food to satisfy the dietary needs of the child cannot be overemphasized.

The typical Western approach to good eating would focus on its nutritional content and calorie intake. In contrast, Korean Medicine has a more comprehensive outlook on eating that weighs diverse coordinates, such as the taste and flavor, color and shape of the food. As an example reflective of this, many highly specific and detailed terms for various medical conditions caused by digestive disorders exist in traditional medicine that in themselves contain the word roots related to "food" or "eating." Included in these are some of the ailments generally considered unrelated to food or eating, like lower back pain, the common cold, headaches, syncope, jaundice, etc. This sheds some light on the etiological paradigm found exclusively in Korean Medicine.

Given the significance that food and eating hold, what does Korean Medicine say on the best way of eating that can guarantee a healthy life for the child?

1) What to eat

Exposure to identical situations can produce vastly different results in different people. For example, the same degree of exhaustion can produce sore throats in some, body aches in others, and indigestion in yet others, depending on the part of the body with the weakest defense. Following suit, not every child will benefit equally from a specific food. Rather, it would be wiser to place children in different categories according to their vulnerabilities and provide them with the complementary food that can strengthen their defense mechanisms.

(1) Children with respiratory problems

There are children who, despite being otherwise healthy with none of the common problems of indigestion, bad bowel habits, or restless sleep, are mysteriously prone to the common cold, sometimes catching as many as six colds a year. This penchant is often discovered in pale-skinned, skinny children with long, swanny necks who cough from the mildest of winds or even from eating cold dish. These kids will benefit from food or tea made with balloonflower, welsh onion, schizandra fruit, walnut, or peach. The balloonflower root, or Platycodi Radix[桔梗], had been used in traditional medicine for is qualities of collecting energy into the lungs and suppressing inflammation of the respiratory organs. Welsh onion, or Allii Radix[葱白], helps lung functions through its ability to disentangle and facilitate the vertical flow of energy and to release energy from the inside. Schizandra fruit, or Schizandrae Fructus[五味子], and walnut are known to gather lung energy and alleviate breathing difficulties, and peach can be used for lung diseases in general. Other remedies include forest hiking or a dry towel scrub that can effect the skin (the skin in traditional medicine is closely related to the lungs).

(2) Children with digestive problems

There are children who, after a full year, still vomit all the time, stormed with erratic bouts of stomachaches and suspicious bowel sounds sprinkled in between. They rotate between diarrhea and constipation, being blighted with an especially weak digestion. Their unfortunate conditions also interfere with natural growth, causing them to fall below average height and weight for their age group. Korean Medicine prescribes for them medicaments and foods that can help digestion and introduce warmth into the internal system, such as mandarine, jujube fruit, glutinous rice, malt, or curled mallow. Mandarine can boost the digestive system and facilitate digestion. The mandarine peel can be processed into a herbal medicine, Citri Pericarpium[橘皮], widely used for indigestion. The fibrous white peels, which can perversely induce dyspepsia, should be discarded. Jujube fruit helps digestive functions and soothes the stomach. However, it can cause dyspepsia if eaten uncooked. The sweet tasting glutinous rice[糯米] is used more commonly for digestive disorders than its non-glutinous counterpart. Glutinous rice porridge can allay abdominal pain and diarrhea. Malt[麥芽], known for its salubrious effects on the digestive system, is used widely in Korea to concoct a special traditional rice drink. Curled mallow[葵] also strengthens the digestive system, usually eaten marinated or used in making soup.

(3) "Fainthearted" children

The children said to be deficient in "heart energy" are depicted in Korean Medicine as being jumpy and easily scared, apt to whimper prior to sleep, unable to get peaceful slumber, prone to nightmares and sleepwalking. Erythruria or thickened, darkened urine can be observed among these children, a sign traditional medicine attributes to a deficiency of "heart energy" and an excess of "heart heat." Some attempt to remedy this by scaring the child deliberately or by exposing the kid to scary scenes for "adaptation." However, this will only exacerbate the problem. Some of the effective ways of counteracting this condition include medicational treatment that can "hearten" the rattly child, such as lotus root, Ixeris dentata lettuce, or red bean. Lotus root[藕節] has cardiotonic effects that soothes the mind and facilitates the flow of "heart energy." The lotus roots and seeds have been used traditionally in Korean Medicine for its strong medicinal effects. Ixeris dentata lettuce[苦菜] has a bitter taste which in traditional medicine is deemed to have cooling properties that can extinguish the excessive "heart heat" and thus bring forth its tranquilizing effects. Red bean[赤小豆], which in the Korean culture had been used for its purported powers against dark magic and evil spirits arising from its redness of hue, is known in medical circles for its ability to send out excessive "heart heat" through urine.

(4) Children with urogenital problems

Children with urogenital problems are marked by frequent urination or urinary incontinence. They are often behind in their potty training, wetting the bed into older years, typically waking up in the mornings with swollen eyelids. We recommend bamboo salt, chestnut, black soybean, sea cucumber, or black domestic fowl (*Gallus gallus var. domesticus*). Bamboo salt, a special salt using bamboo in the roasting procedure, burrows the bamboo's "dampness-eliminating" qualities and its cool nature, ideal for use in the kitchen as a replacement of the common salt. Chestnut can fortify the "kidney energy," attested by the passage in 『Dong-ui bogam』 quoting the benefits of eating roasted chestnuts on daily basis. The color black is said to be related to the kidney in Korean Medicine, perhaps the reason why many black foods are known to strengthen the "kidney energy." Included in theses black foods are black soybean, sea cucumber, and black domestic fowl (*Gallus gallus var. domesticus*).

2) How to eat right

How to eat can carry greater import than what to eat. Indeed, we are not just what we eat, we are *how* we eat.

First, eat in moderate portions. 『Dongeui Bogam』 explains food as one of the factors causing internal disorders, termed "Dietary irregularities飲食傷," or "eating disorder." It puts forth that both excessive hunger and excessive fullness can lead to this malady. Generally speaking, overeating presents more problems than starvation in the modern society. When more food is taken in than the digestive system can muster, blood rushes to the overloaded GI tract, resulting in general decay of body functions and the immune system, which in turn can lead to heightened susceptibility to illnesses.

Second, meals should be pleasant and relaxed. A section in 『Dongeui Bogam』 quotes that "a crying child must be soothed before milking" and enumerates the list of pathological conditions that can arise from forcing milk on an agitated baby. Scolding the child for not eating or milking a baby crying for reasons other than hunger can bring undesirable results.

Third, choose the time of the day for feeding. Eating habits are known to play an important role in regulating biological cycles of the body. Similar to the biorhythm theory, Korean Medicine holds a peculiar view on the physical cycles of the body that involve a delicate balance between the waxing and waning of "Yin" and "Yang." In a cursory explanation, in daytime, the Yang gains dominance while the Yin withers down, while the opposite is true in nighttime. The human body, being a microcosm, responds to the macrocosmic cycles of nature and follows a similar course of event. This is why the humans stay active at daytime, in connection with the active qualities of the Yang energy, and rest at nighttime, matching the inactive qualities of the Yin energy. If the cycles go haywire, our physical well-being is bound to be disturbed. Digestive functions are one of the important activities in our body, and eating long after sunset (after the main dinner) can shatter the biological rhythm. Like being deprived of sleep, our body is deprived of rest when it is needed.

3) Pressure point massage that can be used to treat cold

(1) GB20風池

Location: Near the beginning of the hair line in the back of the head, on the concavity slightly lateral to the two thick muscles running in the neck.

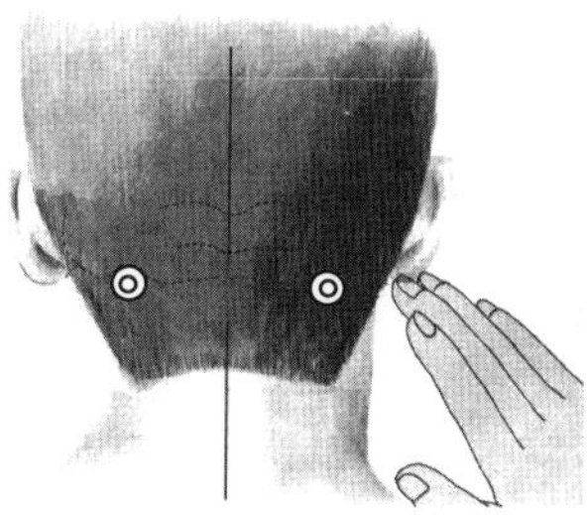

GB20風池

Massage technique: Cover the back of the head with palm, use the thumbs of both hands to massage the pressure points in kneading motions. Alleviates posterior neck pain caused by the cold.

(2) LU1中府

Location: Concavity below the clavicle, lateral to the 2nd rib bone and medial to the shoulder joint.

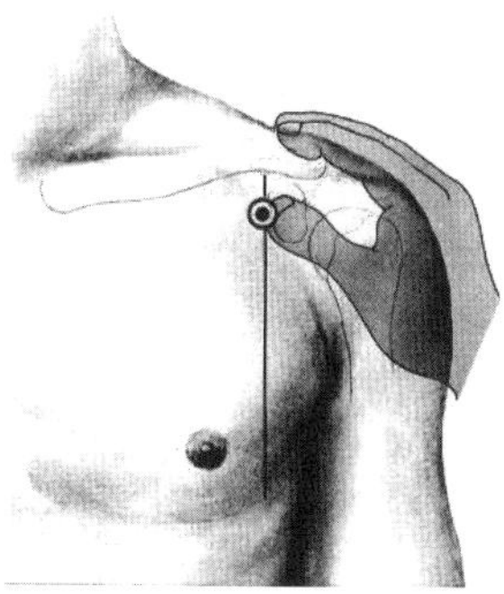

LU1中府

Massage technique: Place the thumbs on the pressure point below the clavicle, wrap the rest of the fingers on the shoulder, press the massage points. Highly effective for respiratory disorders.

(3) BL12風門

Location: Medial to the scapula, flagging the 2nd vertebra hori-

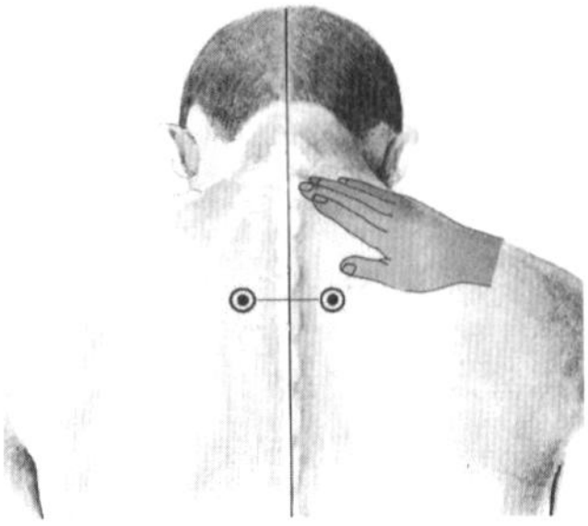

BL12風門

zontally.

Massage technique: Make the patient lie prone, put palms on the back, press the pressure point with both thumbs simultaneously. Effective for respiratory disorders.

3. Constitutionally discriminated recipes

All parents want their children to be healthy, strong, and smart. And this is perhaps why so many resort to herbal tonics and dietary supplement products. But more than anything else, the child will benefit most from homemade meals prepared with recipes matching his/her constitutional type.

Sasang Constitutional Medicine assigns foods to a constitutional type depending on its healthy or unhealthy effects. And naturally, constitutionally discriminated recipes also use constitutionally discriminated foods.

The general rule in preparing meals for a child is to choose proportionately from all food groups. Then, foods and recipes can be chosen that can benefit the child most depending on their compatibility with the child's constitutional type.

1) The Taeyangin child

(1) Seasoned buckwheat paste jelly

Ingredients: 75g buckwheat paste jelly, 5g leek, 1g garlic (chopped), 3g powdered red pepper, 0.5g sesame seed, modicum of black pepper, sugar, and sesame oil

Directions:
1. Cut buckwheat paste jelly into cubes.
2. Take some soy sauce and mix it with chopped leek, garlic, sugar, black pepper, sesame seed, and sesame oil.
3. Mete out the prepared jelly onto plate and serve with sauce on a separate dish or poured onto the jelly.

(2) Common octopus and oyster skewers

Ingredients: 90g common octopus, 40g oyster (unshelled), 30g green onion, 30g oyster mushroom, 5g red pepper paste, 3g sugar, 3g garlic (chopped), modicum of cooking oil

Directions:
1. Scrub the common octopus with salt, wash and parboil in boiling water, cut into edible pieces.
2. Clean oysters in flowing water, parboil in boiling water.
3. Cut green onion and oyster mushrooms into edible sizes.
4. Thread the octopus, oyster, green onion, and mushrooms in rotation on the skewers.
5. Mix red pepper paste, sugar, and chopped garlic together.
6. Brush the sauce thus made onto the skewers and grill lightly, keep adding sauce as it cooks.

(3) Shrimp and broccoli stir-fry

Ingredients: 30g shrimp (peeled), 30g broccoli, 10g leek, 5g soy sauce, 3g cooking oil, 5g starch, modicum of sugar

Directions:
1. Cut broccoli into edible pieces, hack leek into 1.5cm slabs.
2. Boil water, sprinkle in some cooking oil and salt, blanch the broccoli.
3. Mix equal amounts of soy sauce and sugar to

make dressing.
4. Oil the skillet and cook shrimp until pink, mix with broccoli and leek in the same skillet, pour in the prepared seasoning and stir.
5. Stir-fry until the seasoning blends in completely with the vegetables and shrimp, add tints of starch water.
(For better taste: You can make a delicious red pepper paste sauce by mixing the soy sauce with red pepper paste and chili bean sauce. Those ready for spicy adventures may want to use this red pepper paste sauce instead.)

(4) Steamed sea mussel

Ingredients: 80g sea mussel (unshelled), 20g water dropwort, 40g soybean sprout, 3g leek, modicum of garlic, glutinous rice powder, sesame oil, salt, and cooking oil

Directions:
1. Wash and clean dropwort in water, cut 4cm long.
2. Take modicum of water, boil with leek, garlic, ginger, soy sauce, and sugar.
3. Add sea mussel and bring to boil, add soy sprouts and cook.
4. Dissolve glutinous rice powder in water and pour into pot, add dropwort and boil, sprinkle sesame oil and sesame seeds.

2) The Soyangin child

(1) Mung bean pancake

Ingredients: 10g mung bean (peeled), 30g mung bean powder, 40g kimchi (cooked), 10g japanese flower fern, 15g pork (minced), 15g mung bean sprout, modicum of cooking oil and salt

Directions: 1. Mince the kimchi, souse the mung bean in water and grind in blender.
2. Boil mung bean sprouts and fern in water, drain and chop into smaller pieces.
3. Season the minced pork with garlic, salt, and pepper; season the bean sprouts and fern with salt and sesame.
4. Put pork, bean sprouts, and fern in the ground mung bean, add salt.
5. Heat oil in the skillet, ladle the mixture in small morsels, bake into round pancakes.
(For better taste: Mung bean pancakes taste best when cooked slightly thick. Ladle in some more of the mixture onto a halfway cooked pancake and bake until brownish yellow.)

(2) Pork bulgogi

Ingredients: 80g pork, 40g onion, 5g leek, 5g garlic (chopped), 50g pineapple, 10g soy sauce, modicum of ginger, pepper, sesame oil, starch syrup, and soju

Directions: 1. Grind pineapple and onion, mix with soy sauce, sprinkle in ginger, pepper, and garlic to taste. Place pork in the mixture and marinate.
2. Preheat the skillet, stir-fry the marinated pork to desired doneness.
3. Turn off the heat, throw in some leek.

(3) Seasoned flatfish grill

Ingredients: 120g flatfish, 5g cooking oil, modicum of salt, pepper, soy sauce, spring onion, garlic (chopped), sesame oil, and sesame seed

Directions: 1. Clean the flatfish and wash in water, strain and marinate with salt and pepper.
2. Preheat the grill, grill the flatfish to desired done-

ness, brushing the seasoning as it cooks.

(4) Pork and eggplant stir-fry

Ingredients: 80g eggplant, 20g pork (minced), 3g dried red pepper, 5g leek, modicum of garlic, ginger, soy sauce, cooking oil

Directions:
1. Wash eggplant, cut into small, round slices; wash and clean dried red pepper and chop into round slices.
2. Stir-fry the red pepper in oiled skillet, discard them after the spices have spread into the oil, stir-fry garlic and ginger in the same skillet.
3. Place pork in the skillet and stir-fry to desired doneness, add eggplant and cook over high heat.
4. Add soy sauce and sugar when the pork is nearly done, add pepper for spice.
5. Sprinkle the leek just before serving.

3) The Taeeumin child

(1) Sweet potato yogurt salad

Ingredients: 40g sweet potato, 30g apple, 15g mayonnaise, 40g yogurt, modicum of raisin, sugar, mustard sauce, and salt

Directions:
1. Steam the sweet potato and chop into edible pieces.
2. Wash the apple and cut into mouthful sizes.
3. Pour some mayonnaise into the yogurt and mix it with raisin, sugar, mustard sauce, and salt; ladle this mixture on the sweet potato and apple and mix together well.
   (For better taste: Be warned that this recipe dries readily. The dressing should be mixed just before serving.)

(2) Bulgogi

Ingredients: 100g beef, 60g onion, 10g leek, 3g garlic (chopped), 50g pineapple, 10g soy sauce, modicum of ginger, pepper, sesame oil, starch syrup, and soju

Directions:
1. Cut onion into small 0.3cm pieces, marinate beef in prepared seasoning.
2. Oil the skillet and stir-fry the beef.
3. Put in leek and sesame oil when the beef is nearly done and stir-fry for few minutes.

(3) Pumpkin and mushroom stir-fry

Ingredients: 50g white button mushroom, 50g green pumpkin, modicum of salt, cooking oil, and powdered sesame salt

Directions:
1. Wash the pumpkin, cut into crescents.
2. Clean the mushrooms, slice into mushroom shapes or crescents.
3. Stir-fry lightly in preheated skillet, add salt to taste, sprinkle powdered sesame salt before serving.
(For better taste: This recipe can be made more nutritious and tasty, ideal for a kid's snack, by mixing butter with the cooking oil for the stir-fry.)

4) The Soeumin child

(1) Frozen pollack jerky pancake

Ingredients: 60g frozen pollack jerky, 20g egg, 10g flour, modicum of salt, pepper, and cooking oil

Directions:
1. Sprinkle salt and pepper on pollack jerky.
2. Stir eggs well in a bowl; sieve the flour.
3. Smear the pollack with flour thoroughly.
4. Oil the skillet, dip pollack in the egg bowl and

scoop out, fry in skillet until brown.
(For better taste: Low heat should be used to give ample time for cooking.)

(2) Chicken and potato stew

Ingredients: 120g chicken (minced), 40g potato, 20g carrot, 10g soy sauce, modicum of starch syrup, soju, ginger, and sugar

Directions:
1. Wash the minced chicken thoroughly and drain.
2. Mix soy sauce with garlic, sugar, starch syrup, soju, and ginger to make seasoning.
3. Oil the pot and stir-fry the chicken morsels, add potato and carrot.
4. Add the prepared seasoning to the pot and let the stew simmer over low heat until the sauce thickens.
   (For better taste: For the best nutritional content and taste, all ingredients have to be well-cooked on the surface before turning down heat to simmer.)

(3) Samgyetang (Chicken broth with ginseng)

Ingredients: 1 spring chicken, 25g glutinous rice, 10g garlic (peeled), 5g ginseng (not dried), 5g jujube fruit (dried), modicum of pepper

Directions:
1. Do not cut the chicken in the middle, but chip a little at the rear end, discard the guts and clean.
2. Souse glutinous rice in water, let absorb, and drain.
3. Push the rest of the material (glutinous rice, jujube, ginseng, and garlic) into the chicken, tie the chicken legs together to prevent the mixture from splashing out.

4. Pour water in the pot and bring to boil, put in the prepared chicken, boil thoroughly.
5. Place chicken and stock in soup bowl and sprinkle chopped leek, salt, and pepper before serving. (For better taste: To remove the chicken stench, boil the stock with onion or celery.)

(4) Seasoned shepherd's purse

Ingredients: 100g pickpurse, 5g soybean paste, 3g leek, modicum of garlic (chopped), sesame oil, and sesame seed

Directions:
1. Dust and clean the pickpurse, wash in water, blanch in boiling water and drain.
2. Mix soybean paste with chopped garlic, sesame oil, and leek.
3. Toss the mixture and pickpurse in a bowl and mix together throughly until the seasoning blends, sprinkle sesame seed before serving. (For better taste: Clean any dirt or impurities clinging onto the pickpurse roots, pinch in some salt into boiling water and dip in the pickpurse root first, for clearer colors.)

4. Summer health guidelines for children

Our metabolism speeds up to its utmost in the summer, and the nutritional supply needs to match this rate in order for us to spend a healthy and wholesome summer.

In Korean tradition, there are 'canicular days,' the purportedly hottest days of the year, and on these torrid days, our homeostatic system sends out 20-30% more bloodflow to the skin to prevent further rising of the body temperature. This redirection of blood causes loss of appetite, chronic fatigue, and other typical symptoms of being "under the heat." Sometimes, people lose enough sweat to feel exhausted, and the relatively insufficient blood flow to the GI tract causes digestive disturbances and loss of appetite.

The following are some of the guidelines for a healthy summer for the child.

1) Pleasant habitat for pleasant summers

(1) Room temperature

Air conditioning should be strictly controlled so that the room temperature is no lower than 5℃ below the real outside temperature. The cold air should never be directed directly at people, and the windows should be opened at least once an hour to let the air ventilate.

(2) Wash with tepid water

Cold showers after profuse sweating leads to a sudden plunge in body temperature that can cause summer colds. Lukewarm showers are recommended, even on hottest days. The tepid water can wash away the waste materials secreted with the sweat and lower the body temperature gradually without straining our homeostasis.

(3) Avoid cold or iced drinks

Thirst in summer can trigger a craving for icy cold drinks. However, the increase in body temperature causes increased bloodflow to the skin and a subsequent and relative decrease in bloodflow to the viscera. The relatively ischemic conditions of the GI tract exposes the child to increased risks of gastrointestinal malfunctioning. If topped by ingestion of cold beverage, the precarious state can tipple over and affect the bowel movements, leading to diarrhea and associated symptoms. Therefore, children should be made to drink tepid water more frequently to prevent dehydration in the summer.

(4) Solar disinfection

The dampened blankets of children who sweat a lot provide an ideal environment for bacterial growth. Letting the blankets out to dry in the sun can remedy this. The benefits of this procedure, when repeated frequently, is especially evident for those with respiratory allergies who experience dramatic decrease in coughing and phlegm.

(5) Short, within-an-hour daytime naps

Children deprived of sleep become irritable and grouchy, more so after sleepless nights due to the near-lethal nighttime heat waves common in Korea. On these occasions, a nap can galvanize the child, producing more beneficial effects for the more active kids. Indeed, one cannot overemphasize the importance of letting the hyperactive child take a brief nap for a chance to replenish his / her energy.

(6) Children have "cold stomachs"

Normally, children sweat during sleep even when it's not hot, and this is a quite normal thing for kids. This sweating can cause what Korean Medicine calls a "cold stomach." To treat this, cover the stomach well during sleep to prevent heat loss, or massage the belly to facilitate bowel movements, or cook the child foods that are "warm in nature."

2) Ways of spending healthier summers for children of different constitutional type.

(1) Summer health guidelines for the Taeyangin child

The slightest fever can trigger vomiting, and the Taeyangin child, generally a picky eater, is often inflicted with indigestion. Foods that don't comply with his/her constitutional type can induce obesity or constipation, and it is thus important to choose foods compatible with his/her constitutional type. Meat, wheat, and radish are especially unhealthy for the Taeyangin child.

Plain-tasting foods are most healthy. The quantity and frequency of urine is the most significant indicator of health in the Taeyangin child, and a frequent, sufficient urine is a sign of physical comfort. Given this, high glucose foods that concentrates the urine, such as dried persimmon or sugar water, should be avoided.

① Buckwheat noodle: The succulent broth served with the noodle can spice up the child's appetite.
② Green salad: One of the most beneficial foods for the Taeyangin, the green salad can be served when the child loses his/her appetite.
③ Grape juice: The nutritious fruit can replenish energy lost through sweat.
(Persimmon leaf tea is also known to have positive effects, but the strong bitter taste can turn kids off.)

(2) Summer health guidelines for the Soyangin child

The Soyangin child is usually a sturdy kid with a strong digestive system and a voracious appetite. And yet, despite this, their gorging and scrunching eating habit is conducive to dyspepsia or upset stomachs. Therefore, the Soyangin child should be trained to eat slowly in proper, conversational meals. The typical Soyangin is of a "hot constitution" and should avoid the popular nourishing dishes like samgyetang or chicken porridge, which in traditional medical view can generate excessive heat.
① Summer fruits and vegetables like watermelon, cucumber, or oriental sweet melon can serve as ideal foods for the Soyangin that can be stocked for snacks.
② Chinese matrimony vine fruit tea: The chinese matrimony vine fruit is an excellent herb for the Soyangin child. Make some tea out of it and store them in pitchers in the fridge

for easy access.

(3) Summer health guidelines for the Taeeumin child

The Taeeumin child is highly likely to become overweight, constantly plagued with constipation and digestive disorders, and the parents should train the child into appropriate eating styles. Popular summer treats like the soda pop or icecream will add to the risks of gaining weight. We recommend high-protein foods like beef, whereas rich, thick foods are to be avoided. Skimp on the dinners, and never overeat. The Taeeumin child will eat a considerable amount before sensing the first tangs of satiety, and the parents should impose a set limit of food intake to prevent overeating.

① Soybean noodle: The soybean, sometimes called "meat grown on field," is high in nutrition and ideal for the Taeeumin child.
② Milk: Also excellent for the Taeeumin is the soya milk made with soybean.
③ Schizandra fruit punch: The schizandra fruit, with its unique sweet and sour taste, is a herbal fruit most beneficial for the Taeeumin. The delicious schizandra fruit punch is an ideal treat that can swish away the summer heat.
④ Adlay tea: The herb most beneficial for the Taeeumin, adlay can effectively prevent respiratory diseases.

(4) Summer health guidelines for the Soeumin child

The Soeumin child indulges heavily in snacks and pushes away the actual meals. He / she is a picky eater with a fastidious appetite. The typically weak digestive system presents the Soeumin child with incessant bouts of diarrhea, and sweating often leaves him / her more exhausted that expected. Naturally, summer is bound to be a difficult season for the Soeumin. The Soeumin child should not be given iced or cold drinks, even on the hottest days.

Rather, we recommend lukewarm water or a few sips of salt water. Exhausted children should be fed with nourishing foods.

① Chicken porridge: The Soeumin child's frail digestive system will benefit immensely with the chicken, which is known to contain "warm energy."
② Glutinous rice gyeongdan: The special dessert, the Korean version of dumpling-turned-sweet-treat, is compatible with the Soeumin constitution if made with glutinous rice, a grain "warmer" in its nature than its nonglutinous counterpart.
③ Curry: This special condiment can spice up lost appetite.

5. Child growth for different constitutional types

1) General facts on height

There are various factors that can influence child growth. It's the popular belief that children take after parents in most of their traits, including height. However, one study reports that the offspring's height depends only 30% on the parents' height, and the remaining 70% is determined by environmental factors, such as nutrition, exercising, sleep, and stress. Depending on how well these environmental factors are controlled, the child can grow into varying heights in adulthood.

In a different study on the effect of environmental factors on the height of identical twins, twins with the same genetic make-up who were separated in early age and brought up in different environments displayed an average 15cm difference in height. This implies that even individuals with the same DNA material can grow into completely different heights depending on the external factors, such as nutrition, exercising, and sleep.

This hints that children should be provided with a compatible environment in order to grow tall. Balanced nutrition, adequate exercising, positive thinking to counter stress, and deep sleep that can reduce fatigue are some of the prerequisites for activating the growth related genes.

2) Normal growth patterns

The childhood can be divided into four stages according to the growth curve pattern, which includes two growth spurts.

① birth - 2 years: 1st growth spurt
② 2 years - adolescence: slow growth
③ adolescence - 15, 16 years: 2nd growth spurt
④ maturity: decelerated growth

3) Short stature

Normal growth is possible only in healthy conditions. Because of this, height and weight are two important criteria for evaluating the how healthy the child is. Short stature is defined as the state of being below -2.0 SD or third percentile on the height chart for a person of the same age and sex. More simply, it implies that among a group of 100 of the same age and sex, 97 (or more) are taller and only two (or less) are shorter than the child with short stature. Most of the children (led by concerned parents) who visit the medical center on account of "short height" are in fact within normal height range.

4) What causes short stature

Growth is a complex process influenced by diverse factors, hereditary background and environmental surroundings to name a just a few.

Growth retardation can be divided into two groups according to its etiology: primary growth abnormality,[1] related to internal

1 Conditions that can cause primary growth disturbances include skeletal dysplasias (eg.

defects, and secondary growth abnormality,[2] related to external environmental factors. The most common causes found among those seeking for medical help on account of short stature are normal variations of patterns of growth like constitutional growth delay and familial dwarfism.

(1) Familial short stature

Familial short stature is characterized by short height that always remains at a low percentile, height velocity of over 5cm a year, and predicted adult height appropriate to the familial pattern. Puberty begins at normal age, and bone age is normal for chronological age.

(2) Constitutional growth delay

Constitutional delay of growth and puberty refers to a common variation of the growth process in which growth occurs more slowly than average. Bone age is delayed 2-3 years for chronological age, and puberty and pubertal growth spurt occur 2-3 years later than average for both girls and boys. There may be a history of delayed growth and adolescence among the family members. Such small children eventually catch up in later stages as they continue to grow into older age, and their adult height is appropriate to familial pattern.

5) Diagnosis and treatment for short stature

There are several exams presently available in Western medicine to confirm this condition. Some of the simpler tests include hematological test, karyotype test, thyroid function test, urinalysis,

---

achondroplasia), genetic disorders (eg. Turner's syndrome), congenital metabolic disorders, intrauterine growth retardation, and other syndromes associated with short stature.

2 Conditions that can cause secondary growth disturbances include malnutrition, chronic systemic diseases, psychosocial stress, endocrine disorders (eg. growth hormone deficiency, hypothyroidism) and abnormal carbohydrate, lipid, or protein metabolism.

and bone age X-ray. More extensive testing like a growth hormone stimulation test may be required to help confirm growth hormone deficiency.

Short stature is treated according to its causes. If there is a chronic disease disturbing normal growth, the underlying conditions must be treated first. Growth hormone deficiency can benefit from a growth hormone therapy, as can other conditions not associated with growth hormone deficiency, such as Turner's syndrome and chronic renal failure.

6) Formula calculating the child's predicted height

Male offspring: (mother's height + 13cm + father's height)/2
Female offspring: (mother's height - 13cm + father's height)/2
(Predicted height is within 10cm of actual growth)

7) Views on growth in Sasang Constitutional Medicine

Just like your bloodtype, your constitutional type is something you're born with and cannot be changed into another. In order to live healthily, you have to accommodate to your constitutional makeup. Environmental factors should be shifted around to tend to the weaknesses found in your constitutional type. We also recommend constitutionally discriminated regimen that can help promote health.

In Sasang Constitutional Medicine, there is the differentiation of whether a certain foodstuff is beneficial or pernicious for a constitutional type. However, a child should not be denied access to any particular food. Only under pathological conditions, and only if the food is affecting the illness, should the culpable food be removed from the diet. Even then, parents are required to consult a specialist for constitutional diagnosis prior to any final decision.

① Analyzing the child's growth pattern

Most elementary schools have annual spring physical exams, and a quick check on the school records can provide ample information on the growth and developmental progress of the child, which can be used to analyze the chid's growth patterns.

A single, isolated figure on the height chart merely reflects the past growth rate, but the height velocity (cm/yr) calculated from two measurement taken within 6 months from another can reflect the present growth rate. It is up to the parents to learn the normal patterns of growth and to regularly check on their child's growth rate.

Their are many things that can cause short stature, and parents are urged to consult the nearest healthcare practitioner once or twice a year should signs of short stature appear.

Also, children who are at or below third percentile or who grow less than 5cm a year should submit to more a extensive examination by trained specialist.

② Check on the possible factors that might be slowing down growth

Ironically, children are one of the busiest members of our society. After spending hours toiling over schoolbooks, they spend equally long hours for after school curriculums, either in tightly controlled learning centers or in rigorous sport centers. This high-fused agenda naturally entails irregular eating hours that can cause indigestion and other digestive disorders, and the snacks that replace the proper meals are just one of the many threats to wholesome nutrition.

Also, lack of physical activities and insufficient sleep can cause chronic fatigue in the child.

Fatigued children become debilitated and unable to keep up their defense mechanisms, and the weakened immune system is conducive to more frequent and persistent bouts of cold and other chronic conditions.

Parents tend to look out only for the most visible signs in the child, like how tall he/she is. But lying under the cover are the less conspicuous but nevertheless significant factors that can influence growth, such as the eating style, hours of exercising, sleep, diseases, etc. These factors must be taken into account in order to create an ideal environment for unhindered growth. Unnecessary therapeutic measures should not be introduced prematurely.

③ The "Original Symptomology素症"[3] unique to the constitutional type influences child growth

Upon close scrutiny, we can observe in the child the "Original Symptomology素症" unique to the constitutional type, or the group of disorders which a constitutional type has a native and connatural susceptibility to. For example, we hear about children said to have a "particularly weak digestion," or kids who are especially prone to coughing spells in the winter, or others who always seem to have constipation. As we can see, these are everyday conditions that can occur unaccompanied by serious diseases. This set of conditions that a person is prone to is the "Original Symptomology," and it depends largely on the constitutional type of the individual.

The "Original Symptomology" arises from the weakest organ system in a specific constitutional type. In other words, because each type of constitution has fortes and fallibilities in its organ function distribution scheme, or "largeness" and "smallness" in the natural organ energy, the outward manifestation of this inborn imbalance of organ energy takes form of constitutionally differentiated "Original Symptomology." This internal and external imbalance is believed to be the fundamental cause of growth

3 Original Symptomology素症
The "Original Symptomology" is a set of disorders to which the individual of a certain constitutional type has inborn predilection or connatural susceptibility. The "Original Symptomology" can include a wide range of disorders depending on the type of constitution. The "Original Symptomology" arises from constitutional weakness and can naturally develop into or trigger constitutionally specific disorders. The "Original Symptomology" is thus an important criterion in determining the constitutional type of an individual.

disturbances in children. It is closely related to the inherited genetic factors and the past and present environmental factors the child is exposed to, and these factors must be controlled and regulated in order to remedy the growth abnormalities. Indeed, the readjustment of environmental factors is considered the most significant step in treating growth disorders in traditional medicine.

(1) The Taeyangin child

The Taeyangin are of a strong lung system and a weak liver system, having excessive lung energy ("emissive and dispersive energy"; the direction of metabolism moves outward) and deficient liver energy ("absorptive and concentrative energy"; the direction of metabolism centers inward). This gives rise to the "Original Symptomology素症" typical of the Taeyangin, and it is thus essential that this imbalance is corrected in order to restore the pathological conditions found in the Taeyangin.

① "Original Symptomology素症"

- **Insomnia:** A Taeyangin child can become completely hooked up in an activity or event he/she finds interesting, so much as to forgo meals. Parents need to step in and control the child's life gently but firmly. The child yearns to be understood, and upon failure to do so on the part of the parents, the Taeyangin child often experience restless nights and disturbed slumber.
- **Fastidious appetite:** Being choosy eaters, they characteristically refuse food other than traditional Korean rice dishes.
- **Frequent vomiting:** The Taeyangin children often experience gastroesophageal symptoms like vomiting or reflux episodes. The conditions worsen under emotional stress or if the food eaten is found unpalatable. This eating problem is conducive to aberrant bowel habits, sometimes producing pebble-like "goat stools" every 4-5 days. Foamy saliva is another commonly found symptom.

- **Lower limb weakness:** Some report of weakness in lower limb unaccompanied by physical anomalies of the lower back. Some experience nighttime cramps in the calf that occur without any decrease in the muscle mass.

② Encouraging growth through constitutional improvement

The main outlines of treating a Taeyangin child rests on fortifying the "liver energy" to strengthen the internal organs. The commonly used herbal medicines include root bark of araliaceous shrub, kiwi, chinese quince, pine needle, grapevine root, and pine nodular branch. A Taeyangin child will benefit greatly from constitutionally discriminated diets. Plain-tasting food, vegetables, and seafood are among the most compatible foods, whereas precooked food products, wheat products, meat, and dairy products should be eliminated.

- **Easy, homemade herbal potion**

  12g Acanthopanacis Cortex, 8g Chaenomelis Fructus, 8g Pini Nodi Lignum: Place the herbs in glassware or earthenware, pour 400ml of water, leave the herbs in the water for 30min. Bring to boil 3-5min over hight heat, tone down to lowest heat when the water starts to boil. Let simmer for about an hour and reduce the water to 150ml. Imbibe 50ml per portion, three times a day.

(2) The Soyangin child

The Soyangin are of a strong spleen system and a weak kidney system, having excessive spleen energy ("yang and warm energy"; the direction of metabolism ascends upward) and deficient kidney energy ("yin and cool energy"; the direction of metabolism descends downward). This gives rise to the "Original Symptomology[素症]" typical of the Soyangin, and it is thus essential that this imbalance is corrected in order to restore the pathological conditions found in the Soyangin.

① "Original Symptomology素症"

- **Constipation:** Constipation can cause the child to feel congested, heated, and irritable, especially if the child has never had constipation before. If the condition is prolonged, the child can develop bad breath, lose appetite, and his/her face can darken into a somber shade.
- **Inattentiveness and absentmindedness:** The hyperactive kid is incapable of sitting still, and the inattentiveness entails an extremely short attention span.
- **Frequent diarrhea:** The Soyangin child may be attacked by bouts of diarrhea during which the stomach can be icy cold to the touch whereas the hands and feet remain warm. The dehydration can cause thirst, prompting the child to search for cold water. Concentrated, thickened urine can be observed.

② Encouraging growth through constitutional improvement

The main outlines of treating a Soyangin child rests on fortifying the "kidney energy" to introduce a downward flow in the bodily metabolism. The commonly used herbal medicines include Rehmanniae Radix Preparat, Corni Fructus, Poria, and Moutan Cortex.

- **Easy, homemade herbal potion**

  16g Rehmanniae Radix Preparat, 8g Lycii Fructus, 8g Corni Fructus, 4g Alismatis Rhizoma, 4g Poria: Place the herbs in glassware or earthenware, pour 400ml of water, leave the herbs in the water for 30min. Bring to boil 3-5min over hight heat, tone down to lowest heat when the water starts to boil. Let simmer for about an hour and reduce the water to 150ml. Imbibe 50ml per portion, three times a day.

(3) The Taeeumin child

The Taeeumin are of a strong liver system and a weak lung system, having excessive liver energy ("absorptive and concentrative energy"; the direction of metabolism centers inward) and deficient lung energy ("emissive and dispersive energy"; the direction of metabolism moves outward). This gives rise to the "Original Symptomology素症" typical of the Taeeumin, and it is thus essential that this imbalance is corrected in order to restore the pathological conditions found in the Taeeumin.

① "Original Symptomology素症"

- **Overweight:** The Taeeumin are by nature lazy and therefore inactive (thus unable to use up the ingested calories), and are typically gluttonous, traits that can easily lead to obesity.
- **Bowel movements set off immediately by eating:** Though born with a healthy stomach, the rest of their GI tract, the intestines, are relatively weak, and the Taeeumin children are often attacked by involuntary and urgent bowel movements.
- **Excessive sweating:** The Taeeumin are hopelessly intolerant of hot weather, often the ones who sweat profusely from the slightest physical activity. Naturally, they are always thirsty, and all the drinking makes them urinate frequently and in large amounts.

② Encouraging growth through constitutional improvement

The main outlines of treating a Taeeumin child rests on fortifying the "lung energy" to introduce an outward flow in the bodily metabolism. The commonly used herbal medicines include Ephedrae Herba, Coicis Semen, Dioscoreae Rhizoma, Asparagi Radix, Puerariae Radix, and Cervi pantotrichum Cornu.

- **Easy, homemade herbal potion**

  8g Cervi Cornu, 8g Liriopis Tuber, 8g Platycodi Radix, 8g Coicis Semen: Place the herbs in glassware or earthenware, pour 400ml of water, leave the herbs in the water for 30min. Bring to boil 3-5min over hight heat, tone down to lowest heat when the water starts to boil. Let simmer for about an hour and reduce the water to 150ml. Imbibe 50ml per portion, three times a day.

(4) The Soeumin child

The Soeumin are of a strong kidney system and a weak spleen system, having excessive kidney energy ("yin and cool energy"; the direction of metabolism descends downward) and deficient spleen energy ("yang and warm energy"; the direction of metabolism ascends upward). This gives rise to the "Original Symptomology素症" typical of the Soeumin, and it is thus essential that this imbalance is corrected in order to restore the pathological conditions found in the Soeumin.

① "Original Symptomology素症"

- **Indigestion:** A slightest step outside the normal amount of food can trigger digestive disturbances in the Soeumin child, leading to headaches, indigestion, and diarrhea.
- **Cold extremities:** Disrupted digestive system is conducive to cold hands and feet, according to Traditional Korean Medicine. These people are also prone to bouts of nauseous bloating and gurgling bowel sounds.
- **Sweaty extremities:** The Soeumin children typically report of cold sweat around the neck, sweaty hands and feet, and digestive disorders. The stickiness of the sweat can be a source of annoyance for the child.
- **Listless mornings:** In most people, a night's slumber can wash away the pent up fatigue. But a Soeumin child finds mornings especially difficult to deal with, and with choice will sleep through them, skipping breakfast.

② Encouraging growth through constitutional improvement

The main outlines of treating a Soeumin child rests on fortifying the "spleen energy" to introduce an upward flow in the bodily metabolism. The commonly used herbal medicines include Ginseng Radix, Atractylodis macrocephalae Rhizoma, Cinnamomi Cortex, and Astragali Radix.

- **Easy, homemade herbal potion**

  4g Ginseng Radix, 4g Atractylodis macrocephalae Rhizoma,

4g Paeoniae Radix Alba, 4g Glycyrrhizae Radix: Place the herbs in glassware or earthenware, pour 400ml of water, leave the herbs in the water for 30min. Bring to boil 3-5min over hight heat, tone down to lowest heat when the water starts to boil. Let simmer for about an hour and reduce the water to 150ml. Imbibe 50ml per portion, three times a day.

Appendix 1 Constitutional self-diagnosis

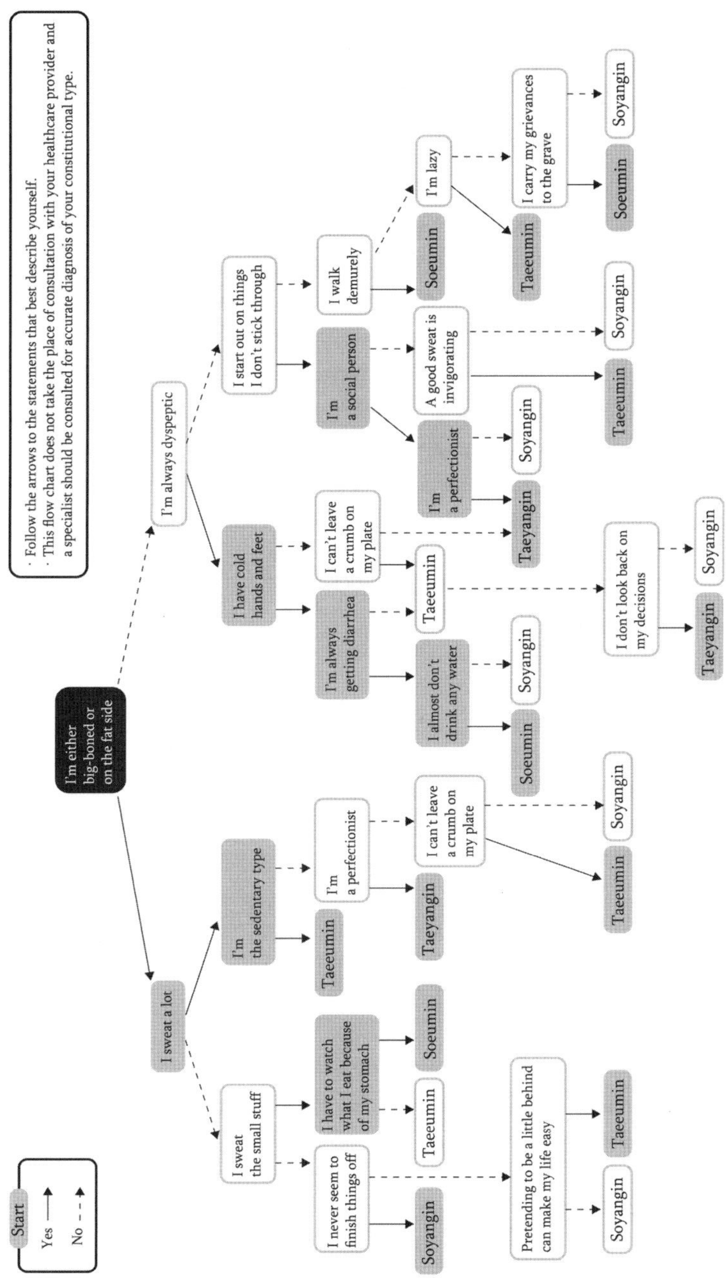

Appendix 2 Beneficial foods for each constitutional type

The Department of Sasang Constitutional Medicine in Kyung Hee Korean Medicine Center has divided the foods mentioned in 『Sallim Gyeongje』, 『Jeungbo Sallim Gyeongje』, 『Imwon Gyeongjeji』, and 『Gyuhap Chongseo』 according to its taste and nature, which were then assigned to one of the four constitutional types depending on its compatibility. Two hundred fifty one of these were selected and placed into 16 categories (grains and grain products, potato and starch, beans and bean products, seeds and nuts, meat and meat products, milk and dairy products, egg, fish and shellfish, seaweeds, fruits, vegetables, mushrooms, condiments, beverage, herbal medicine, others) according to 『2008 Report on Nutrition and Health in Korea』.

Food groups	Food	Taeyangin	Taeeumin	Soyangin	Soeumin
grains and grain products	millet		+*	*	
	hemp		+		
	buckwheat	+*		*	×
	nonglutinous millet	+*		+*	
	nonglutinous rice				+
	wheat		+		×
	flour		+		×
	barley	*		+*	×
	bread (flour)		+		
	Indian millet		+		
	powdered malt			+	
	corn		+	+*	+*
	Job's tears		+		
	hulled millet			+*	+*
	glutinous millet				+
	wheat flour				+
	barnyard millet				

(+ beneficial food, × harmful food, * edible food)

Food groups	Food	Taeyangin	Taeeumin	Soyangin	Soeumin
	brown rice		+	+	
	black rice				*
potato and starch	potato				+
	sweet potato		+*	*	
	taro	*	+	*	
beans and bean products	haricot		+		
	green bean	*		+*	×
	soy bean		+		
	bean-curd		+*	*	
	red bean	*		+*	
seeds and nuts	hazelnuts	*	+		+*
	acorn		*		*
	perilla japonica		+*		*
	peanuts		+		
	chestnut		+		
	ginkgo		+		
	pine-nuts		+		
	sesame	*		+*	
	sunflower seed		*		*
	walnut	+*	+*		*
meat and meat products	goose	×		+	
	pheasant	×			+
	deer	×		×	+
	chicken	×	×	×	+
	pork	×	×	+	×
	quail	×			+
	snake	×			+
	beef	×	+		
	lamb	×			+
	goat	×	×	×	+
	duck	×		+	
	sparrow	×			+
	turkey	×			+
	rabbit	×		+*	+*
milk and dairy products	butter		+		
	milk		+*	*	×
	cheese		+		
eggs	chicken egg		×	+*	*
	duck egg			+	

(+ beneficial food, × harmful food, * edible food)

Food groups	Food	Taeyangin	Taeeumin	Soyangin	Soeumin
fish and shellfish	mullet	*		+ *	
	flatfish			+	+
	crayfish	*		+ *	
	cutlass fish		+ *		+ *
	crab	+ *	+	+ *	
	mackerel	*	+	*	×
	whale		+		
	oyster	+ *	*	+ *	
	Ammodytes personatus Girard				+
	saury	*	+	*	×
	small octopus	+ *		+ *	
	flatfish	*		*	
	sea bass				+
	codfish		+ *		+ *
	sandfish				+
	sea bream	*		*	+
	ascidian			+	
	catfish		*		+ *
	anchovy		*		+ *
	pollack		+		
	Alaska pollack		*		+ *
	octopus	+		*	
	loach		*		+ *
	Styela clava	*		*	+
	brown croaker		+ *		+ *
	Tapes philippinarum	*		*	
	eel		+ *		+ *
	Japanese Icefish		+		+
	butter fish	*		*	
	puffer	*		+ *	
	carp	+ *			*
	Japanese maker		*		*
	shrimp	+ *		+ *	
	echinoidea	*		*	
	spiny turban shell	+ *	+ *	*	
	salmon		+		
	flathead mullet				+
	mandarin fish				+

(+ beneficial food, x harmful food, * edible food)

Food groups	Food	Taeyangin	Taeeumin	Soyangin	Soeumin
	chum salmon		+		
	atka mackerel		*		
	squid	+*	*	+*	×
	snail	*	+	+*	
	sweetfish				+
	carp		+*	+*	
	snapping turtle	*		+*	
	eel	*		*	+
	abalone	+		+	
	sardine			+	
	yellow corvina		+		+
	filefish			*	
	tuna		+*		*
	herring		*	+*	
	sea cucumber	+*	*	+*	
	skate	*		*	
	mussel	+*		*	+*
seaweeds	laver	+*	+*	*	*
	sea tangle	*	+*	*	+*
	brown seaweed	*	+*	*	*
	Bengal isinglass		+		
	jellyfish		+		
fruits	persimmon	*		*	
	tangerine				+
	the fruit of Actinidia arguta	+*		*	
	Chinese date				+
	strawberry	*		+*	
	lemon		*		+*
	wild grapes	+			
	melon	*		+*	
	plum		+		
	papaya	+			
	banana	*	*	+*	×
	Chinese pear	*	+*	*	
	cherry		*		*
	peach				+
	apple		×		+
	wild strawberry			+	
	apricot	+	+*	*	

(+ beneficial food, × harmful food, * edible food)

Food groups	Food	Taeyangin	Taeeumin	Soyangin	Soeumin
	pomegranate		*		+*
	watermelon	*		+*	×
	mulberry		+		
	oranges	+*			*
	citron	*		*	+
	plum	*	+	*	
	kiwi	*		*	
	pineapple	+		+	
	grapes	+*		*	
vegetables	eggplant	*		+*	
	leaf mustard		*		×
	Korean lettuce			+	+*
	Asmunda japonica		*		
	fernbrake		+*		*
	red pepper		*	×	+*
	sugar beet	*		*	+*
	sesame leaf				*
	pickpurse				+
	wild garlic		*		+*
	carrot		+*		+*
	Codonopis lanceolata		+		
	Chinese bellflower		+		
	aralia shoots		+		
	garlic		×	×	+
	butterbur		+		
	radish		+*		*
	parsley				+
	Chinese cabbage	*		+*	
	scallion		*		+*
	Chinese pepper		*		*
	lettuce	+*		+*	
	ginger		×	×	+
	celery	+		+	
	purslane			+	
	sponge gourd	*		+*	
	green bean sprout			*	
	watershield plant	+			
	spinach	*		*	+
	mugwort				+

(+ beneficial food, × harmful food, * edible food)

Food groups	Food	Taeyangin	Taeeumin	Soyangin	Soeumin
	crown daisy				+
	sowthistle	*		+*	
	marsh mallow			+	
	cabbage				+
	onion		*		+*
	lotus root		+		
	young radish	*		+*	
	cucumber	*		+*	
	burdock	*		+*	
	bamboo shoot		*	+*	
	plantain	*		+*	
	wild aster		+*		*
	bean sprouts	*	+	*	
	tomato	+*			
	shallot		*	×	
	green pepper		*		
	pumpkin		+	+	
mushrooms	shiitake mushroom	*	+*	*	
	agaric mushroom	*	+	*	
	Auricularia (auricula-judae)		+	*	+*
	Phellinus linteus		+		+*
	Tricholoma matsutake	*	+*	*	
	Flammulina velutipes	*	+	*	
condiment	mustard		*		
	soybean paste	*	+	*	*
	sugar		+		*
	salt				+*
	vinegar		*		+
	sesame oil			+	*
	fermented soybeans	*		*	+
	curry			×	+
	black pepper		×	×	
beverage	green tea	*		+*	×
	soymilk		+*	*	*
	rice wine				+
	beer			+	
	so-ju (distilled liquor)		+*		*

(+ beneficial food, × harmful food, * edible food)

Food groups	Food	Taeyangin	Taeeumin	Soyangin	Soeumin
	goat milk				+
	yogurt		+		
	coffee	+	×		
	cocoa		*		
herbal medicine	licorice				+
	Cassiae Torae Semen	*		+*	+
	cinnamon				+
	persimmon	*		*	
	Amorphophalus konjac	*		*	
	the fruit of Chinese matrimony vine	*		+*	*
	young antlers of a deer		+*		+
	Angelicae gigantis Radix				+
	wax gourd		*	*	*
	Eucommia ulmoides		*		+
	royal jelly			×	+
	Trapa japonica	*		*	*
	Epimedium koreanum		*		
	R. Radix preparata	*		+*	
	squalene	*	+	*	
	aloe	*		+*	
	Ganoderma Lucidum	*		+*	
	Acanthopanax	+			
	Schizandra chinensis Baillon		+		
	Euphoria longana		+		+
	Cistanches Herba			+	
	motherwort				+
	ginseng				+
	Siegesbeckia glabrescens	*		*	
	Astragalus embranaceus				+
others	tree of heaven	*		*	
	brine				+
	liver oil		+		
	Shank Bone Soup		+		
	mum		+		
	whitlow grass	*		*	
	honey		×	×	+
	sundrop oil			+	

(+ beneficial food, × harmful food, * edible food)

Food groups	Food	Taeyangin	Taeeumin	Soyangin	Soeumin
	snail	*		*	
	chrysalis		*		*
	loquat	*	*	*	
	ice creams			+	×
	beef soup with rice		+		
	pine needles	+			
	castor bean			*	*
	malt				+*
	lacquer				+
	safflower			+	
	solar salts			+	
	kuzu vine		+		
	safflower oil			+	

(+ beneficial food, × harmful food, * edible food)

Appendix 3 Constitutional diet plans for diabetics

Taeyangin

meal	day	1	2	3	4	5	6	7	8	9	10
Breakfast	"rice"	"rice"- cereals, millet	"rice"- cereals, barley	"rice"- german millet	"rice"- barley	"rice"- german millet	"rice"- barley	"rice"- german millet	"rice"- barley	"rice"- german millet	"rice"- barley
	"soup"	"potstew"- oyster	"soup"- beanpaste mallow	"soup"- shellfish	"soup"- beanpaste mudsnail	"soup"- hard-shelled mussel	"soup"- beanpaste blue crab	"soup"- dried codfish	"soup"- beanpaste shepherd's purse	"soup"- beanpaste mallow	"soup"- dried pollack
	"meat dish"	"gui"- butterfish	"gui"- yellow carvenia	"jorim"- frozen pollack	"gui"- flatfish	"jorim"- butterfish radish	"boggum"- octopus with red pepper	"muchim"- sliced turbo with mustard	"boggum"- octopus with red pepper	"jorim"- mackerel radish	"gui"- squid
	"side dish"	smothered egg	"jorim"- small crab	"namul"- eggplant	"namul"- chard	"muchim"- dried codfish	smothered egg	"jjim"- hard shell mussel	"namul"- herbs	toasted laver	"boggum"- sea cucumber vegetables
	"vegetable dish"	"muchim"- raw lettuce	"muchim"- cucumber	"muchim"- raw cabbage	"namul"- shepherd's purse	"namul"- sesame leaf	"namul"- young radish red pepper paste	"muchim"- cucumber	"namul"- amaranthus	"jjim"- sesame leaf	"namul"- cabbage
	"kimchi"	"kimchi"- young radish	"kimchi"	"kimchi"- young radish	"kimchi"- watery radish	"kimchi"	"kimchi"- young radish	"kimchi"	"kimchi"	"kimchi"- young radish	"kimchi"- young radish

Lunch	"rice"	"rice"-cereals, barley	"rice"-cereals, millet	"rice"	"rice"-german millet	panfried "rice"-shrimp	"rice"-german millet	"rice"-barley	"rice"-german millet	"rice"-barley	"rice"-german millet
	"soup"	"soup"-beanpaste mudsnail	"soup"-blue crab red pepper paste	"soup"-beanpaste cabbage	"soup"-beanpaste spinach	"soup"-egg, green onion	"soup"-dried pollack	"soup"-beanpaste short-necked clam	"soup"-blue crab red pepper paste	"soup"-egg, green onion	"soup"-cabbage
	"meat dish"	"boggum"-squid with red pepper	"jeon"-oyster coated with egg	"gui"-sliced duck	"jorim"-mackerel pike kimchi	octopus, oyster on a skewer / sauce	"muchim"-smothered skate / red pepper sauce	fried dried pollack / sesame sauce	"boggum"-shrimp ball	"gui"-clam	"gui"-mackerel pike
	"side dish"	toasted laver	"muchim"-raw cabbage	parboiled cabbage / bean paste sauce	"muchim"-smothered cockle	"muchim"-raw lettuce	scrambled egg	buckwheat jelly / soy sauce	"boggum"-squash	salad	"jorim"-cockle
	"vegetable dish"	song-I mushroom salad	"boggum"-squash	"namul"-herbs	"namul"-green onion	"gui"-song-I mushroom	"namul"-cabbage	raw brown seaweed / vinegar, hot paste	"muchim"-raw cabbage	tomato	"namul"-green onion
	"kimchi"	"kimchi"	"kimchi"-young radish	"kimchi"	"kimchi"-young radish	"kimchi"	white "kimchi"-not powdered red pepper	"kimchi"	white "kimchi"-not powdered red pepper	"kimchi"	"kimchi"
Dinner	"rice"	"rice"	"rice"-barley	"rice"-german millet	"rice"-barley	"rice"-german millet	"rice"-barley	"rice"-german millet	"rice"-barley	"rice"-german millet	"rice"-barley
	"soup"	"soup"-hard-shelled mussel	"soup"-beanpaste	"soup"-fishjelly	"soup"-melanian snail	"soup"-beanpaste	"potstew"-oyster	"soup"-vegetables	"soup"-squid, radish	"soup"-shellfish	"soup"-dumpling

			young radish			mallow		sea-cucumber			
Dinner	"meat dish"	"jorim"-herring radish	"gui"-flatfish	"gui"-porgy	"gui"-sardine	smothered octopus	"boggum"-pork with red pepper paste	"jjim"-porgy	"muchim"-oyster	"jjim"-skate, bean sprout / red pepper paste	"jeon"-squash
	"side dish"	"boggum"-shrimp	"jorim"-burdock	egg roll	"jorim"-potato	"jorim"-small crab	raw lettuce / paste sauce	egg roll	"boggum"-potato pyogo mushroom	"jeon"-oyster coated with egg	"namul"-young radish bean-paste
	"vegetable dish"	"namul"-amaranthus	parboiled cabbage / bean paste sauce	"muchim"-brown swaweed	"muchim"-wild rocambole	"namul"-amaranthus	"muchim"-dropwort	"muchim"-laver	"namul"-spinach	"namul"-eggplant	"muchim"-cucumber
	"kimchi"	"kimchi"	"kimchi"	"kimchi"-watery radish	"kimchi"	"kimchi"-young radish	"kimchi"	"kimchi"	"kimchi"-young radish	"kimchi"	white "kimchi"-not powdered red pepper

"gui"-food rosated [grilled] with seasonings
"kimchi"-a spicy pickle or mixture containing cabbage, onions, and something fish, variously seasoned, as with garlic, horseradish, red pepper
"guk"-soup, "jorim"-food boiled down in soy with spices, "pulgogi"-broiled beef preserved with sauce and seasoning
"muchim"-combine together with added ingredients and seasonings, "jjim"-heat with steam in sauce with spice
"namul"-combine together with parboiled vegatable and seasonings, "boggum"-panbroiled [fry in oil] food with spices, "jeon"-panfried food coated with flour

Soyangin

	meal \ day	1	2	3	4	5	6	7	8	9	10
Breakfast	“rice”	“rice”-cereals, barley	“rice”-cereals, barley	“rice”	“rice”-cereals, millet	“rice”-red bean	“rice”-corn	“rice”-barley	“rice”-barley	“rice”-german millet	“rice”-german millet
	“soup”	“soup”-cabbage	“soup”-beanpaste mudsnail	“casserole”-kimchi	“potstew”-soybean curd, salted shrimp	“soup”-dried pollack	“soup”-mushroom	“soup”-beanpaste	“soup”-beanpaste	“soup”-beef	“potstew”-fermented
	“meat dish”	“gui”-flatfish	“boggum”-pork, kimchi	“gui”-sardine	“jorim”-pork	“boggum”-sea cucumber vegetables	“gui”-porgy	“jjim”-ribs	“gui”-mackerel	“gui”-flatfish	“jjim”-shank of beef
	“side dish”	“jorim”-beef, egg	“jjim”-squash	“jorim”-small crab	“jjim”-sesame leaf	“boggum”-pork, cucumber	“namul”-eggplant	“gui”-soybean curd	“boggum”-squash	“jorim”-beef, egg	“namul”-cabbage
	“vegetable dish”	“muchim”-cucumber	“muchim”-raw lettuce	“muchim”-herb	“namul”-thin strips of radish	“namul”-young radish bean-paste	celery	“namul”-amaranthus	“jjim”-duck egg	“namul”-thin strips of radish	toasted laver
	“kimchi”	“kimchi”-herbs	white “kimchi”-not powdered red pepper	“kimchi”	“kimchi”	white “kimchi”-not powdered red pepper	“kimchi”-young radish	“kimchi”	“kimchi”-cucumber	“kimchi”	“kimchi”-herbs
Lunch	“rice”	“rice”-cereals, millet	“rice”-barley	“rice”-cereals, barley	“rice”	“rice”-cereals, millet	“rice”-german millet	“rice”-red bean	“rice”-barley	“rice”-barley	“rice”-barley
	“soup”	“soup”-blue	“soup”-rib	“soup”-	“casserole”-	“potstew”-	“soup”-	“soup”-dried	“soup”-	“soup”-	“soup”-dried

		crab red pepper paste		beanpaste mallow	octopus hot paste	squid	shellfish	pollack	beanpaste short-necked clam	dumpling	pollack
Lunch	"meat dish"	"gui"-lobster with salt	"muchim"-oyster	"boggum"-pork red pepper paste	"jjim"-duck egg	"gui"-herring	pork cutlet	"pulgogi"-duck	"jjim"-blue crab / hot pepper paste	"jeon"-squash	"rice"-bean sprout
	"side dish"	"muchim"-young radish	toasted laver	parboiled cabbage / bean paste sauce	"jorim"-burdock	"jeon"-meatball	salad	"namul"-mushroom	"namul"-eggplant	raw carrot, cucumber	"boggum"-sea cucumber vegetables
	"vegetable dish"	5 strawberrys	"namul"-eggpl ant	"muchim"-raw radish	"muchim"-raw herbs	"muchim"-brown seaweed with vinegar	watermelon	"muchim"-wild rocambole cucumber	fried brown seaweed	"muchim"-raw cabbage	"muchim"-raw lettuce
	"kimchi"	"kimchi"	"kimchi"-cucumber	"kimchi"	"kimchi"-cucumber	"kimchi"	white "kimchi"-not powdered red pepper	"kimchi"-young radish	"kimchi"	white "kimchi"-not powdered red pepper	"kimchi"
Dinner	"rice"	"rice"-corn	"rice"-cereals, millet	"rice"-red bean	"rice"-cereals, barley	"rice"-barley	"rice"-barley	"rice"-german millet	"rice"-german millet	"rice"-red bean	"rice"-german millet
	"soup"	"potstew"-soybean curd (tofu), beanpaste	"soup"-shellfish	"soup"-beef brown seaweed	"soup"-cabbage	"potstew"-fermented soybean	"potstew"-roe of a pollack	"soup"-bean sprout	"soup"-beanpaste (japanese)	"soup"-mushroom	"soup"-marine product / red pepper paste
	"meat dish"	"jorim" -herring	"boggum"-squid	"jorim"-flatfish	"jjim"-ribs	sliced boiled pork in spices	"jjim"-tofu, egg	"gui"-lobster with salt	"boggum"-pork with red	"gui"-eel	"jeon"-squash

		radish							pepper paste		
Dinner	"side dish"	raw sea-squirt / vinegar, hot paste sauce	egg roll	"muchim"-laver	"jeon"-oyster	raw cabbage / bean paste sauce	"jeon"-corn	toasted laver	raw lettuce / paste sauce	"namul"-mallow bean-paste	"namul"-beansprout
	"vegetable dish"	celery	"boggum"-mixed dish of vegetables, lmanna lichen	"boggum"-squash	"boggum"-bamboo shoot	"muchim"-raw radish oyster	"namul"-thin strips of radish	"namul"-amaranthus	"namul"-spinach	"boggum"-bamboo shoot	"muchim"-raw radish oyster
	"kimchi"	"kimchi" -young radish	"kimchi"	"kimchi"	"kimchi"	white "kimchi"-not powdered red pepper	"kimchi"-cucumber	"kimchi"	"kimchi"	"kimchi"	"kimchi"

"gui"-food rosated [grilled] with seasonings
"kimchi"-a spicy pickle or mixture containing cabbage, onions, and something fish, variously seasoned, as with garlic, horseradish, red pepper
"guk"-soup, "jorim"-food boiled down in soy with spices, "pulgogi"-broiled beef preserved with sauce and seasoning
"muchim"-combine together with added ingredients and seasonings, "jjim"-heat with steam in sauce with spice
"namul"-combine together with parboiled vegatable and seasonings, "boggum"-panbroiled [fry in oil] food with spices, "jeon"-panfried food coated with flour

Soeumin

	meal \ day	1	2	3	4	5	6	7	8	9	10
Breakfast	"rice"	"rice"-glutinous millet	"rice"-brown rice	"rice"-cereals, glutinous rice	"rice"-cereals, millet	"rice"-chinese millet	"rice"-brown rice	"rice"-cereals, glutinous rice	"rice"-brown rice	"rice"-cereals, glutinous rice	"rice"-brown rice
	"soup"	"soup"-potato	"soup"-mushroom	"soup"-cabbage	"soup"-kimchi	"soup"-mugwort	"soup"-dried pollack	"soup"-beanpaste shepherd's purse	peppery "potstew"-tofu not yet pressed	"soup"-frozen pollack	"soup"-kimchi, soybean curd not yet pressed
	"meat dish"	"gui"-flatfish	"gui"-yellow carvenia	"jorim"-chicken	"gui"-croaker	"jorim"-tuna radish	"boggum"-mixed dish of green pepper, meat	"pulgogi"	"gui"-yellow carvenia	"jorim"-beef	"boggum"-frozen pollack
	"side dish"	"jorim"-quailegg green pepper	"jorim"-cockle	"boggum"-anchovy in redpepper paste	"jorim"-lotus root	"muchim"-parboiled butterbur stalk, penilla	smothered egg	"jorim"-tofu	"boggum"-dried icefish	"jeon"-lotus root	"jorim"-potato seasoned with hot pepper
	"vegetable dish"	"muchim"-parboiled spinach	fried seatangle	"muchim"-raw scallion	"namul"-sesame leaf	"namul"-fernbrake	"muchim"-parboiled herbs with beanpaste	"namul"-crowndaisy	"jjim"-green pepper	"namul"-sesame leaf	"muchim"-parboiled herbs
	"kimchi"	"kimchi"	"kimchi"	"kimchi"	"kimchi"	"kimchi"	"kimchi"	"kimchi"	"kimchi"	"kimchi"	"kimchi"
Lunch	"rice"	"rice"-brown rice	"rice"-chinese millet	"rice"-brown rice	"rice"-cereals, glutinous rice	"rice"-cereals, millet	"rice"-chinese millet	"rice"-glutinous	"rice"-chinese millet	"rice"-glutinous	"rice"-chinese millet

								millet, millet		millet	
Lunch	"soup"	"soup"-codfish peppery seasonings	"soup"-loach	"soup"-chicken peppery seasonings	"soup"-beanpaste squash	"soup"-beanpaste wild rocambole	"soup"-thick beef, noodles	"soup"-beef brown seaweed	"soup"-cabbage	"casserole"-kimchi	"soup"-bean sprout
	"meat dish"	"boggum"-chopped chicken vegetables	"muchim"-dried codfish	"jeon"-sliced frozen pollack	"pulgogi"	"boggum"-turkey	"gui"-eel	"gui"-hairtail	"pulgogi"-chicken	"gui"-herring	"gui"-croaker
	"side dish"	"muchim"-hardshelled mussel, radish	"jorim"-peanut	"boggum"-pyogo mushroom	"boggum"-potato vegetables	"muchim"-jellyfish mustard	"muchim"-raw scallion	"jorim"-sliced codfish	"boggum"-anchovy green pepper	"namul"-herbs	"jorim"-chicken leg seasoned with curry
	"vegetable dish"	"muchim"-parboiled shepherd's purse	"jjim"-green pepper	"muchim"-edible shoots of a fatsia with vinegar	"muchim"-raw herbs	"muchim"-green onion	apple salad	"namul"-beansprout	"muchim"-roots of balloonflower	"boggum"-potato vegetables	"jorim"-pepper
	"kimchi"	"kimchi"	"kimchi"	"kimchi"-leaf-mustard	"kimchi"-leaf-mustard	"kimchi"	"kimchi"	"kimchi"	"kimchi"	"kimchi"	"kimchi"
Dinner	"rice"	"rice"-cereals, glutinous rice	"rice"-cereals, millet	"rice"-chinese millet	"rice"-brown rice	"rice"-cereals, glutinous rice	"rice"-cereals, millet	"rice"-corn	"rice"-cereals, millet	"rice"-chinese millet	"rice"-hulled millet
	"soup"	"soup"-beanpaste wild rocambole	"soup"-beef brown seaweed	"potstew"-tofu, potato	"soup"-potato sea tangle	"soup"-codfish	"soup"-taro	"soup"-hard-shelled mussel	"soup"-potato	"potstew"-wall eye pollack	"soup"-thin strips of radish
	"meat	"pulgogi"-	"jorim"-hairtail	"gui"-perch	"gui"-mackerel	panbroiled	"potstew"-	"jorim"-	"jorim"-flatfish	"boggum"-	"boggum"-beef

Dinner	dish”	turkey	radish		pike	chicken with soy sauce	chicken spicy seaning	mackerel pike kimchi	radish	mixed dish of green pepper, meat	with hot pepper paste
	“side dish”	“boggum”-dried icefish	smothered cabbage / beanpaste sauce	soft tofu / soy sauce	“jorim”-quailegg	“jeon”-tofu	fried seatangle	“gui”-codonops is lanceolata	“jorim”-cockle	fried fowl / mustard	“muchim”-raw herbs
	“vegetable dish”	“muchim”-raw dropwort	“muchim”-wild rocambole	“namul”-beansprout	“muchim”-parboiled shepherd's purse	“namul”-spinach	“muchim”-radish	“muchim”-brown seaweed with vinegar	“boggum”-pyogo mushroom	“muchim”-wild rocambole	brown seaweed / vinegar, hot pepper paste
	“kimchi”	“kimchi”-leaf-mustard	“kimchi”	“kimchi”	“kimchi”	“kimchi”	“kimchi”	“kimchi”	“kimchi”	“kimchi”	“kimchi”

“gui”-food rosated [grilled] with seasonings
“kimchi”-a spicy pickle or mixture containing cabbage, onions, and something fish, variously seasoned, as with garlic, horseradish, red pepper
“guk”-soup, “jorim”-food boiled down in soy with spices, “pulgogi”-broiled beef preserved with sauce and seasoning
“muchim”-combine together with added ingredients and seasonings, “jjim”-heat with steam in sauce with spice
“namul”-combine together with parboiled vegatable and seasonings, “boggum”-panbroiled [fry in oil] food with spices, “jeon”-panfried food coated with flour

Taeumin

meal	meal / day	1	2	3	4	5	6	7	8	9	10
Breakfast	“rice”	“rice”-brown rice	“rice”-kidney bean	“rice”-chinese millet	“rice”-brown rice	“rice”-kaoliang	“rice”-adlay	“rice”-kidney bean	“rice”-chinese millet	“rice”-brown rice	“rice”-kaoliang
	“soup”	“soup”-thin strips of radish	“soup”-beanpaste mallow	“soup”-beef brown seaweed	“soup”-beanpaste mudsnail	“soup”-thin strips of radish, perilla	“soup”-thin strips of squash	“casserole”-pyogo mushroom	“soup”-beanpaste wild rocambole	“soup”-panfried radish	“soup”-agaric mushroom
	“meat dish”	“pulgogi”	“jorim”-beef, pepper	spicy beef-rolled in a cabbage leaf	“gui”-minced rib	“pulgogi”	“jjim”-shank of beef	“jorim”-meatball	beef-rolled with glutinous paper	“gui”-minced meat	“jjim”-shank of beef
	“side dish”	“gui”-hairtail	toasted laver	“jorim”-peanut	“gui”-croaker	“gui”-flatfish	“boggum”-anchovy pepper	“gui”-dried pollack	“boggum”-dried icefish	“jorim”-lotus root	“jorim”-codfish radish
	“vegetable dish”	“muchim”-parboiled herbs	“muchim”-brown seaweed	“muchim”-raw radish oyster	“namul”-crowndaisy	“namul”-parboiled radish	“namul”-spinach	“namul”-fernbrake	“muchim”-parboiled herbs with beanpaste	“boggum”-bamboo sprout	“muchim”-parboiled butterbur stalk, penilla
	“kimchi”	“kimchi”	“kimchi”	“kimchi”	“kimchi”	“kimchi”	“kimchi”	“kimchi”	“kimchi”	“kimchi”	“kimchi”
Lunch	“rice”	“rice”-chinese millet	“rice”-brown rice	“rice”-kaoliang	“rice”-adlay	“rice”-kidney bean	“rice”-chinese millet	“rice”-brown rice	“rice”-kaoliang	“rice”-adlay	“rice”-kidney bean
	“soup”	“potstew”-soybean curd (tofu)	“casserole”-mushroom	“soup”-frozen pollack peppery seasonings	“potstew”-tofu squash	“soup”-beef radish	“soup”-thick beef	“soup”-bean sprout	“soup”-frozen pollack	“soup”-taro	“potstew”-soybean curd (tofu)

Lunch	"meat dish"	"jeon"-meat ball	"gui"-tofu	"muchim"-sliced boiled meat	"gui"-spit	oven"gui"-mackerel	"gui"-spainsh mackerel	sliced boiled meat	sliced raw beef	"jeon"-tofu, beef	"gui"-spit
	"side dish"	"jorim"-mackerel radish	"muchim"-jelly fish mustard	"boggum"-agaric bamboo sprout	"jorim"-hairtail radish	"jorim"-tofu	"joen"-squash	raw sea tangle	"michim"-jellyfish mustard	"gui"-croaker	"jorim"-butterfish radish
	"vegetable dish"	"namul"-beansprout	"namul"-squash	"namul"-fernbrake	"muchim"-codonopsis lanceolata	"muchim"-parboiled herbs	raw brown seaweed / vinegar, hot paste	raw carrot, sweet potato	"namul"-flowering fern	"namul"-beansprout	"namul"-spinach
	"kimchi"	"kimchi"	white "kimchi"-not powdered red pepper	"kimchi"	white "kimchi"-not powdered red pepper	"kimchi"-cucumber	"kimchi"	"kimchi"-wrapped in a large cabbage leaf like a bundle	"kimchi"-young radish	"kimchi"	"kimchi"
		glutinous rice cake coated with bean flour, milk	biscuit, milk	sweet potato 1/2, milk	bread, milk						
Dinner	"rice"	"rice"-kaoliang	"rice"-adlay	"rice"-kidney bean	"rice"-chinese millet	"rice"-chestnut, ginkgo nut	"rice"-kaoliang	"rice"-adlay	"rice"-kidney bean	"rice"-chinese millet	"rice"-brown rice
	"soup"	"soup"-shank of beef	"soup"-bean sprout	"soup"-beanpaste spinach	"potstew"-ham sausage, beef	"soup"-beanpaste shepherd's purse	"soup"-beanpaste mallow	"soup"-anchovy brown seaweed	"soup"-beanpaste spinach	"soup"-beef brown seaweed	"soup"-beanpaste mudsnail

Dinner	"meat dish"	"jeon"-sliced frozen pollack	"boggum"-beef pimento	beef-rolled with glutinous paper	"boggum"-mixed dish of vegetables, meat, noodle	boiled meat in five spices	"jorim"-beef	"gui"-yellow carvenia	"pulgogi"	spicy beef-rolled in a cabbage leaf	"jorim"-meat ball
	"side dish"	"boggum"-pyogo mushroom	"gui"-dried pollack	"jorim"-tuna, radish	"boggum"-flowering fern	"boggum"-f rozen pollack	"gui"-hairtail	"gui"-codonopsis lanceolata	"gui"-tuna	smoked salmon / sauce	"gui"-yellow carvenia
	"vegetable dish"	"muchim"-roots of balloonflower	raw sea tangle red pepper, vinegar	raw carrot, sweet potato	agaric with vinegar, red pepper paste sauce	"namul"-beansprout	"muchim"-roots of balloonflower	pyogu mushroom soybeanpaste sauce	"jjim"-soybeanpaste sauce	"muchim"-parbroiled dropwort	"namul"-scallion
	"kimchi"	"kimchi"-radish cabbage	"kimchi"	"kimchi"	"kimchi"-cucumber	"kimchi"	"kimchi"-young radish	"kimchi"	"kimchi"	"kimchi"	"kimchi"

"gui"-food rosated [grilled] with seasonings
"kimchi"-a spicy pickle or mixture containing cabbage, onions, and something fish, variously seasoned, as with garlic, horseradish, red pepper
"guk"-soup, "jorim"-food boiled down in soy with spices, "pulgogi"-broiled beef preserved with sauce and seasoning
"muchim"-combine together with added ingredients and seasonings, "jjim"-heat with steam in sauce with spice
"namul"-combine together with parboiled vegatable and seasonings, "boggum"-panbroiled [fry in oil] food with spices, "jeon"-panfried food coated with flour

Appendix 4 Beneficial herbal medicine for each constitutional type

1) Soeumin

Aconitum carmichaeli Debx 附子

Aconitum carmichaeli Debx 川烏

Agastache rugosa (Fisch. Et Mey.) O. Kuntze 藿香

Allium fistulosum Linne 葱白

Allium sativum L. 大蒜

Alpinia officinarum Hance 良薑

Alpinia oxyphylla MIQ. 益智

Amomum cardamomum L. 白蔻

Amomum xanthioides WALK. ex BAK. 砂仁

Angelica sinensis (OLIV.) DIELS 當歸

Aquillaria agallocha RoxB. 沈香

Areca catechu L. 腹皮

Areca catechu L. 檳郎

Arisaema erubescens (Wall.) Schott 南星

Artemisia capillaris Thunberg 茵蔯

Artemisia princeps Pampan (=*A. asiatica* Nakai, *A. montana* Pampan) 艾葉

Asarum sieboldii MIQ., *Asarum maculatum* NAKAI 細辛

Astragalus membranaceus Bunge 黃芪

Atractylodes chinensis Koidz., *Atractylodes lancea* DC 白朮

Atractylodes chinensis Koidz., *Atractylodes lancea* DC 蒼朮

Caproelus pygargus 獐肝

Carthamus tinctorius L. 紅花

Cinnamomum cassia (Presl.) Blume 桂皮

Cinnamomum loureirii NEES 桂枝

Cinnamomum loureirii NEES 肉桂

Citrus unshiu Markovich 陳皮

Citrus unshiu Markovich 青皮

Cnidium officinale Makino 川芎

Corydalis turtschaninovii Besser 玄胡索

Crataegus pinnatifida Bunge 山査

Croton tiglium L. 巴豆

Curcuma Zedoaria Roscoe 蓬朮

Cyperus rotundus L. 香附

Echinochloa crus-galli (L.) P. Beauv. 稷米

Elsholtzia ciliata (Thunb.) Hylander 香薷

Eugenia aromatica (L.) Merr 丁香

Faeces Trogopterorum 五靈

Foeniculum vulgare Mill. 小茴

Gallus domesticus Brisson 雄鷄

Glycyrrhiza uralensis Fisher 甘草

Halloysitum Rubrum 赤石脂

Honey (Stone honey) 石蜜

Illicium verum Hooker fil. 大茴

Inula helenium L. 木香

Leonurus sibiricus Linne 益母草

Limonite Fe4O3(OH)6 or 2Fe2O3 · 3H2O. 禹餘粮

Lindera strychinifolia (Sieb. et Zucc.) F. Vill. 烏藥

Liquidambar orientalis MILL. 蘇合油

Magnolia officinale REHDER et WILS., *Magnolia obovata* THUNB. 厚朴

MASSA MEDICATA FERMENTATA 神麯

Melia azedarach Linne 練子

Myristica fragrans Houtt. 肉蔻

Oraza sativa L. var. *glutinosa* Matsumura 糯米

Orixa japonica Thunb 常山

Paeonia lactiflor Pall. var. *hotensis* Makino 白芍

Paeonia lactiflor Pall. var. *hotensis* Makino 赤芍

Panax ginseng C. A. Meyer 人蔘

Papaver somniferum L. 罌粟殼

Perilla frutescens Britton var. *acuta* Kudo 蘇子

Perilla frutescens Britton var. *acuta* Kudo 紫蘇

Phasianus colchicus 雉肉

Pinellia ternata (Thunb.) Makino 半夏

Piper longum L. 華茇

Piper nigrum, L. 胡椒

Placenta Hominis 紫河車

Pleuropterus multiflorus Turcz 何首烏

Poncirus trifoliata Rafin 枳殼

Poncirus trifoliata Rafin 枳實

Prunus persica Batsch 桃仁

Psoralea corylifolia Linne 破故紙

Santalum album L. 檀香

Setaria italica (L.) P. Beauv. 粟米

Solanum tuberosum subsp. tuberosum 甘藷

Sorbus commixta Hedl 丁公藤

Sparganium erectum Linne 三稜

Styrax benzoin Dryand. *S. tonkinensis* (Pier.) Craib 安息香

Terminalia chebula Retzius 訶子

Tetradium ruticarpum (Benth.) Hartley (=*Euodia (Evodia) rutaecarpa*

Benth.) 吳茱

Trionyx sisnensis, Carapax Amydae 鱉甲

Zanthoxylum schinifolium S. et Z. 川椒

Zingiber officinale Rosc. 乾薑

Zingiber officinale Rosc. 生薑

Zizyphus jujuba Miller var. *inermis* Rehder (=*Z. jujuba* Mill.) 大棗

2) Taeeumin

Acorus gramineus Soland. *Acorus gramineus* Soland. var. variegatus HORT. 菖蒲

Ampelopsis japonica (Thunb.) Makino 白斂

Angelica dahurica Benth. et Hooker 白芷

Asparagus cochinchinensis (Lour.) Merrill 天門冬

Boiled Egg of domestic fowl 熟鷄卵

Bos taurus domesticus Gmelin 牛肉

Bos taurus domesticus Gmelin 牛黃

Castanea crenata Sieb. et Zucc. 栗子

Cimicifuga heracleifolia Kom. *Cimicifuga simplex* Wormsk. *Aruncus dioicus* (Walt.) Fern. *var. kamtschaticus* Hara 升麻

Coix lacryma-jobi Linne var. mayuen (Romain) Stapf. *Coix lacryma-jobi* Linne 薏苡

Cornu Cervi Parvum 鹿角膠

Cornu Cervi Parvum 鹿茸

Curcuma longa Linne 鬱金

Cyprinus carpio 鯉膽

Dioscorea tenuipes Fr. et Sav. 薯蕷

Dryobalanops aromatica Gaertn.F 龍腦

Ephedra sinica Stapf, *Ephedra equisetina* bunge 麻黃

Euphoria longan (Lour.) Steud. 龍眼

Fossilia Ossis Mastodi 龍骨

Ginkgo biloba L. 白果

Gleditsia sinensis Lam., *G. japonica* Miq. var. *koraiensis* (Nak.) Nakai 皂角

Glycine max Merr. 大豆黃卷

Hinoceros unicornis 犀角

Holotrichia 蠐螬

Hrysanthemum morifolium Ramat. 菊花

Ilanthus altissima (Mill.) Swingle for. altissima 樗根

Ligusticum Sinense OLIV. 藁本

Liriope platyphylla WANG et TANG, Mondo japonica var. umbrosus MAXIM, Liriope spicata LOUR. 麥門

Malus asiatica Nakai. 陵苓

Morus alba Linne, Morus bombycis KOIDZ, Morus dissecta Nakai 桑皮

Moschus moschiferus L. 麝香

Nelumbo nucifera Gaertn 蓮肉

Physalis alkekengi Linne var. *franchetii* (Masters) Hort. 酸漿

Plantago asiatica Linne, *Plantago depressa* Willd, *Plantago major* Linne var. *japonica* (Fr. et Sav.) Miyabe, *Plantago camtschatica* Chrm 車前葉

Platycodon grandiflorum (Jacq.) A. DC 桔梗

Polygala tenuifolia Willdenow 遠志

Pueraria thunbergiana (Sieb. et Zucc) Benth 葛根

Pyrus pyrifolia (Burm. fil.) Nakai var. culta (Makino) Nakai 梨

Raphanus sativus var. *hortensis* for. *acanthiformis* Makino 蘿蔔子

Raphanus sativus var. *hortensis* for. *acanthiformis* Makino 蘿葍

Raphanus sativus var. *hortensis* for. *acanthiformis* Makino 萊菔根

Rheum officinale Baillon, Rheum palmatum Linne var. palmatum, Rheum undulatum Linne 大黃

Runus armeniaca LINNE 杏仁

Runus mume Sieb. Et Zucc 烏梅

Schizandra chinensis (TURCZ.) BAILL. *Schizandra nigra* MAXIM *Kadsura japonica* (THUNB.) DUNAL 五味子

Scutellaria baicalensis Georgi 黃芩

Selenarctos thibetanus G. CUVIER, *Ursus arctos* L. 熊膽

Spirodela polyrhiza (Linne) Schleid., *Lemna perpusilla* Torrey 浮萍

Thuja orientalis LINNE 栢子

Thuja orientalis LINNE, *Thuja occidentalis* LINNE 側栢葉

Typha latifolia LINNE, *Typha angustata* BORY et CHAUB. *Typha orientalis* PRESL 蒲黃

Ucumis melo L. 瓜蒂

Uisqualis indica LINNE 使君

Ussilago farfara L. 款花

Viviparidae (river snail) 田螺

Zizyphus jujuba Mill. 酸棗

3) Soyangin

Akebia quinata (THUNB.) DECNE. *Akebia Quinata* DECNE. var. *polyphylla* NAKAI 木通

Alisma canaliculatum A. BR. et BOUCHE. 澤瀉

Aloe vera LINNE, *Aloe arborescens* AIV. 蘆薈

Anemarrhena asphodeloides Bunge 知母

Angelica decursiva (MIQ.) FR. et SAV. 前胡

Arctium lappa L. 牛蒡子

Boswellia carterii Birdwood 乳香

Bupleurum falcatum LINNE, *Bupleurum longiradiatum* TURCZ. *Bupleurum euphorbioides* NAKAI., *Bupleurum latissimum* NAKAI. 柴胡

Cistanche deserticola Y. C. Ma 蓯蓉

Commiphora myrrha Engl. 沒藥

Coptis chinensis Franch, *Coptis japonica* (Thunb.) Makino 黃連

Cornus officinalis Sieb. Et Zuce 山茱

Corylus heterophylla var. *thunbergii* Bl 榛子

Cucumis melo L. 甜瓜

Egg of domestic fowl 生鷄卵

Euphorbia kansui T. N. Liou et T. P. Wang 甘遂

Forsythia koreana Nakai, *Forsythia saxatilis* Nakai, *Abeliophyllum distichum* Nakai 連翹

Gardenia jasminoides Ellis for. *grandiflora* Makino 梔子

Globefish / Tetrodontida 河豚

Gypsum Fibrosum (CaSO4 · 2H2O) 石膏

Heracleum moellendorffii Hance 獨活

Hordeum vulgare Linne var. *hexastichon* Aschers. 麥芽

Hordum sativum Jess. var. *hexaastichon* Hack. 大麥

Ledebouriella seseloides (Hoffm.) Wolff., *Ledebouriella divaricata* (Turcz.) 防風

Lonicera japonica Thunb. *L. hypoglaus* Miq. *L. confusa* DC. *L. dasystyla* Rehd. *Lonicera japonica* Thunb., *Lonicera japonica* Thunb. var. *repens* (Sieb.) Rehder for. 忍冬

Chinensis Watsonhara, *Lonicera sempervirens* Linne 金銀花

Lycium chinense Mill. *Lycium barbatum* Linne 枸杞

Lycium chinense Mill., *L. barbarum* L. 地骨皮

MASSA MEDICATA FERMENTATA 神麯

Mentha arvensis Linne var. *piperascens* Malinv 薄荷

Morus alba Linne 黑桑椹

Ostericum koreanum (Maxim) Kitagawa, *Angelica japonica* A. Gray 羌活

Ostrea gigas Thunb 石花

Paeonia suffruticosa Andrews 牧丹

Phaseolus calcaratus Roxb 赤豆

Phaseolus radiatus Linne 綠豆

Phellodendron amurense Rupr., *Phellodendron molle* Nakai, *Phellodendron insulare* Nakai 黃栢

Phyllostachys nigra Munro var. *henonis* Stapf 竹瀝

Phyllostachys nigra Munro var. *henonis* Stapf 竹茹

Plantago asiatica Linne, *Plantago depressa* Willd, *Plantago major* Linne var. *japonica* (Fr. et Sav.) Miyabe, *Plantago camtschatica* Chrm 車前

Platycodon grandiflorum (Jacq) A. Dc. 瓜蔞仁

Polyporus umbellatus (Pers.) Fries 豬苓

Poria cocos (Schw.) Wolf 茯苓

Poria cocos Wolf 茯神

Rehmannia glutinosa (Gaertn.) Libosch., *R. glutinosa* Libosch. f. *hueichingensis* (Chao et Schih) Hsiao 生地

Rehmannia glutinosa (Gaertn.) Libosch., *R. glutinosa* Libosch. f. *hueichingensis* (Chao et Schih) Hsiao 熟知

Rubus coreanus Miq., *Rubus oldhamii* Miq., *Rubus ribesioideus* Matsumura. *Rubus idaeus* Linne var. *concolor* Nakai, *Rubus phoenicolasius* Maxim. *Rubus sorbifolius* Maxim, *Rubus schizostylus* Lev. 覆盆子

Schizonepeta tenuifolia Briquet var. *japonica* Kitagawa 荊芥

Scrophularia buergeriana Miq., *Scrophularia koraiensis* Nakai, *Scrophularia kakudensis* Fr. 玄蔘

Sodii Sulfas 芒硝 (major component ; Na2SO4 · 10H2O 96-98%, MgSO4, CaSO4, NaCl etc)

Sodium Borate (natural Borax) 硼砂

Sophora flavescens Arr., *Echinosophora koreensis* Nakai 苦蔘

Stichopus japonicus Selenka 海蔘

Stomachicum Galli Corium 鷄內金

Sus scrofa var. domesticus 猪肉

Talcum (3MgO.4SiO2.H2O) 滑石

Trichosanthes kirilowii MAXIM. 天花粉

Triticum satuurum Lamarck (*T. aestivum* L., *T. vulgare* Vill. 小麥

4) Taeyangin

Acanthopanax sessiliflorus (RUPR. et MAX.) SEEM, *Acanthopanax seoulense* NAKAI, *Acanthopanax chiisanensis* NAKAI, *Acanthopanax sieboldianum* MAKINO, *Acanthopanax koreanum* NAKAI 五加皮

Actinidia arguta Planch, *A. rufa* Planch, *A. chinensis* Planch 獼猴桃

Brasenia schreberi J. F. Gmel. 蓴

Carassius auratus 鯽魚

Chaenomeles sinensis KOEHNE, *Chaenomeles speciosa* (SWEET) NAKAI 木瓜

Diospyros kaki Thunb 柿子

Fagopyrum esculentum Moench 蕎麥

Gentiana scabra BUNGE var. *buergeri* (MIQ.) MAX. *Gentiana triflora* PALL. *Gentiana uchiyamai* NAKAI, *Gentiana jamesii* HEMSL, *Gentiana axillariflora* LEV. et VNT. var. *coreana* (NAKAI) KUDO 草龍膽

Imperata cylindrica (LINNE) BEAUV. var *koenigii* (RETZ.) DURAND et SCHINZ 茅根

Phragmites communis Trin 蘆根

Pinus densiflora Sieb. Et Zucc 松花

Pinus densiflora Sieb. Et Zucc 青松節

Prunus armeniaca Linne 杏肉

Prunus tomentosa Thunb 櫻桃

Vitis vinifera L. 葡萄

Appendix 5 Standard Terminologies on Sasang Constitutional Medicine

Term	Chinese	Definition / Description	Ref
abidance	居處	one of the Four Humanly Affairs; the settlement (establishment) that provides the primary necessities of life (food, clothing, shelter) that serves as the most basic foundation for the human living	WHO-IST
activity-mind-body-matter	事心身物	Refers to the basic factors of Sasang Constitutional Medicine to understand all nature.	
administration	事務	one of the Four Humanly Affairs; the diverse forms of social phenomena (political, economical) including occupation, education, work, etc.	
aggregation	黨與	one of the Four Humanly Affairs; akin relationships similar to blood relationships	
ample urination; unrestricted urination	小便旺多	indication of healthiness in the Taeyang type; related to assessment of urination, a process carried out in the Liver Affiliation, in order to assess the level of inspirational and concentrative energy of the liver in the Taeyang type, who is born with a "large" lung system and a "small" liver system	
aplomb	威儀	the confidence and stateliness naturally acquired while learning the basic skills necessary to adapt to the surroundings	WHO-IST
appearance and manner of speech	容貌詞氣	an active method of observing the physical attributes in order to discern the constitutional type of a given individual; includes the facial features and manner of speech	
ascending force	上升之力	the driving force responsible for using the thick dregs of the pool of Jin; the ascending force of the esophagus is used to take the thick dregs of the pool of Jin and fortify the esophagus itself	

Term	Chinese	Definition / Description	Ref
association	交遇	one of the Four Humanly Affairs; human relationships in social contexts	
attentiveness (diligence)	勤幹	one of the tonic factors, avoidance of intoxication	
avarice (covetousness)	貪慾	one of the toxic factors, related to opulence	
bi-bo-tan-nuo; four vices	鄙薄貪懦	the state of having discarded the Four Virtues (benevolence, righteousness, propriety, and wisdom) and having succumbed to the Covetous Desires	
bone	骨	the end product generated when the thick dregs of the pool of Jeong are processed by the driving force of the foot (the flexing and withstanding force)	
brisk perspiration; unobstructed perspiration	汗液通暢	indication of healthiness in the Taeeum type; related to assessment of perspiration, a process carried out in the Lung Affiliation, in order to assess the level of expirational and dispersive energy of the lung in the Taeeum type, who is born with a "large" liver system and a "small" lung system	
chest	胸襟	the body part pertinent to the mid-upper burn-center	
chest and axilla	胸腋	thorax; a site pertinent to the spleen	
collecting and gathering force	能收之力	the driving force responsible for using the thick dregs of the pool of Mak; the collecting and gathering force of the hand is used to process the thick dregs of the pool of Mak and form muscle	
comfortable digestion; unstrained digestion	飮食善化	indication of healthiness in the Soeum type; related to assessment of digestion, which reflects the sufficient existence of the spleen energy, in the Soeum type, who is born with a "large" kidney system and a "small" spleen system	
Conspectus; Compendium; Appendix; Addendum; Supplement	泛論	passages inserted at the end of the discussion of each constitutional type, introduces the medicinal formulae for miscellaneous diseases, the general approach for constitutional pathologies, and the preservative methods based on constitutional differentiation	
constant (tensional) mindset; constant (tensional)	恒心	the mental propensity (under tension) to always advance, lift, remain, and stay, related to the Seong-manifestation,	WHO-IST

Term	Chinese	Definition / Description	Ref
disposition		and the mental propensity (under tension) to always lean toward masculinity, feminity, conquering, and guarding, related to the Jeong-manifestation	
constitution	體質; 稟賦	The characteristics of an individual, which include structural and functional characteristics, temperament, adaptability to environmental changes, and susceptibility to disease. A person's constitution is determined in part by genetics and is in part acquired, thus it is relatively stable.	WHO-IST
constitutional medicine	體質醫學	The branch of medicine, which is based on assessment of an individual's constitution and involves physiology, pathology, diagnosis, treatment, and maintenance of health.	
consuming force	消導之力	the driving force responsible for using the thick dregs of the pool of Yu; the consuming force of the small intestine is used to take the thick dregs of the pool of Yu and fortify the small intestine itself	
containing force	停畜之力	the driving force responsible for using the thick dregs of the pool of Go; the containing force of the stomach is used to take the thick dregs of the pool of Go and fortify the stomach itself	
cowardice	怯心	constant cowardice; the general mindset of the Taeeum type	
cranium	頭腦	the site where the clear energy of the pool of Jin, after coming out by the driving force of the ear (force driven by hearing) to be thence converted in to Sin, goes in to form the pool of Ni; the body part pertinent to the upper burn-center	
crest (nape)	腦傾	the body part pertinent to the upper burn-center	
descending force	下降之力	the driving force responsible for using the thick dregs of the pool of Aek; the descending force of the large intestine is used to take the thick dregs of the pool of Aek and fortify the large intestine itself	

Term	Chinese	Definition / Description	Ref
discomfit	不安定之心	constant discomfit; the general mindset of the Soeum type	
disregard (dismissal)	怠慢	the element connected to premature death in Lee Jema's paradigm of self-regulation	
drawing in and sending out of nutrient material; nutrient metabolism	出納水穀;水穀代謝	the water-food metabolism carried out in the Kidney and Spleen systems (storeroom for nutrient metabolism); related to the physiology, pathology, and therapeutics of SY and SE types (in the superior-inferior dimension)	WHO-IST
dread	懼心	constant dread; the general mindset of the Soyang type	
earthly space	地方	one of the Four Heavenly Frames; the foundation for existence, highly similar to the concept of nature (nature-related element)	
easy defecation; unhindered defecation	大便善通	indication of healthiness in the Soyang type; related to assessment of defecation, which reflects the sufficient existence of the kidney energy (in light of the fact that the Spleen Affiliation is the gateway of hot energy and the Kidney Affiliation is the gateway of cold energy) in the Soyang type, who is born with a "large" spleen system and a "small" kidney system	
embracing momentum	包勢	the physical comportment reflective of the shelling and embracing property of the spleen	
equilibrium and harmony / equilibrium and moderation	中和/中節	the unbreached state where sorrow-anger-joy-pleasure is not yet set in motion (equilibrium) and the just state where the sorrow-anger-joy-pleasure set in motion arrives at in appropriateness (harmony)	
esophagus	胃脘	one of the Four Bu Organs, the site where the warm energy is formed when the lighter component of the hot energy ascends	
Esophagus Cold-based Exterior Cold disease	胃脘受寒表寒病	pathological condition caused by "smallness" of the lung system and the resultant deficiency in the expirational and dispersive energy of the lung; can be divided into the Baechu-Pyo symptomatology and the Wiwan-han	WHO-IST

Term	Chinese	Definition / Description	Ref
		symptomatology	
exhale-dispersing qi	呼散之氣	The qi that promotes health for Taeeum type.	WHO-IST
exhaling and inhaling of energy and fluid; energy metabolism	呼吸氣液; 氣液代謝	the energy-fluid metabolism carried out in the Liver and Lung systems (gateway for energy metabolism); related to the physiology, pathology, and therapeutics of TY and TE types (in the interior-exterior dimension)	WHO-IST
experience	經綸	the accumulation of rich experience through collection of diverse experience and knowledge	
exterior disease; exterior-based disease	表病	the pathology in which the Seong damages the exterior energy	
exterior-interior combined (compound) disease	表裏俱病	the pathology in which the Seong and Jeong damages the exterior and interior energy	
Exterior-origin Lower back disease (the same term in WHO-ICTM)	外感腰脊病	represented by paraparesis; the excess dispersive energy of the lung and the deficient concentrative energy of the liver together produce the Taeyangin physical property of superior strongness and inferior weakness, for which reason the lower back (pertinent to the liver system) is susceptible to exopathogen, providing basis for pathology localized at the lower back; the upper body is intact, while the lower body becomes loose and disabled, making it difficult for the patient to ambulate	WHO-IST
flesh	肉	the end product generated when the thick dregs of the pool of Hyeol are processed by the driving force of the waist (the loosening and releasing force)	
flexing and withstanding force	屈强之力	the driving force responsible for using the thick dregs of the pool of Jeong; the flexing and withstanding force of the foot is used to process the thick dregs of the pool of Jeong and form bone	
force driven by hearing-seeing-smelling-tasting	聽視嗅味之力	the driving force influencing the clear energy of the Four Anterior Pools, is related to the ear-eye-nose-mouth; the power of observing the Heavenly frame; responsible for taking out the clear energy of the Four Anterior Pools to fill the Four Burn-Centers and form	

Term	Chinese	Definition / Description	Ref
		Sin-Gi-Hyeol-Jeong, after which it is then sent to the head-should-waist-hip to form the Four Posterior Pools	
force driven by sorrow-anger-joy-pleasure	哀怒喜樂之力	the driving force influencing the clear juice of the Four Posterior Pools, is related to the lung-spleen-liver-kidney; the power of conducting the Humanly Affair; responsible for taking in the clear juice of the Four Posterior Pools, which is used to nourish its source and preserve/activate/condense the Four Anterior Pools	
forgetfulness	健忘	the major pathology that can be seen when the dread of the Soyangin escalates to greater dread, a severe Soyangin pathology	
foundational work	資業	that which was given to man as an element of the Course (命) when the universe was brought to life; the basis for projecting the Path; the specific method of doing the Acting	
fouraffiliations	四黨	a collective terminology for the metabolic and circulatory pathway of the energy derived from water-food, the interactive relationship of the affiliated organs and tissues in each burn-center that is centered around the controlling Jang organ	
four anterior pools	前四海	the accumulation in the anterior part of the body of the substances created from the biological transformations within the Four Affiliations	
four basic quaternary elements; heaven-human-knowing-acting / heaven- human-nature-course	天人知行 / 天人性命	the basic elements representing the universe and the human (extrinsic element—Heavenly Loom, Humanly Affair; intrinsic element—Knowing, Acting)	
four Bu organs	四腑	the site where the four energies (warm-hot-cool-cold) are formed as the water-food passes through; the site where the energy derived from water-food is separated according to its clarity or turbidity by the mechanism of ascent and descent in the superior-inferior dimension	

Term	Chinese	Definition / Description	Ref
four burn-centers	四焦	the site of vital biological transformations that take place within the Four Affiliations, extends to the concept of the Four Affiliations	
four inattentive vices	驕矜伐夸	the four inattentions that can be overcome through proper Knowing; the covetous greed one can easily succumb to during the process of Knowing	
four indolent vices	奪侈懶竊	the four indolences that can be overcome through proper Acting; the covetous greed one can easily succumb to during the process of Acting	
four posterior pools	後四海	the accumulation in the posterior part of the body of the substances created from the biological transformations within the Four Affiliations	
four ruinous causes	酒色財權	the temptation one can fall prey to if one's knowledge and action are improper and thusly induce inattention and indolence, can trigger pathological conditions; inattention in knowledge and indolence in action (negligence, indulgence, avarice, and prejudice) causes one to be susceptible to the four ruinous causes that drains health, whereas valuing the good and scorning the evil (attentiveness, simplicity, knowledgeability, and vigilance) can restore health	
gluteal area; infra-spinal area	腰脊下	the site related to the kidney; the posterior body part related to the lower-most burn-center	
greed; covetous desire	心慾	antithesis of the Four Virtues of benevolence, righteousness, propriety, and wisdom	
head-shoulder-waist-hip	頭肩腰臀	the body part involved in doing the Acting	
heavenly frame	天機	the environmental factors surrounding the human being; the collective term for all environmental factors that always surround the human being and influence the human life (objective, environmental element)	
heavenly time	天時	one of the Four Heavenly Frames; the changes over time or chronological property (time-related element)	

Term	Chinese	Definition / Description	Ref
humanly activity	人事	the features of the human life; the collective term for the specific elements in the human life (objective, social element)	
humanly ethics	人倫	one of the Four Heavenly Frames; the ethics, morality, or tribal/racial qualities that are formed amid human relationships (humanity-related element)	
inattentive mind	邪心	rejection of the inattentive mind is the foundation for doing the "knowing"; includes arrogance, assertiveness, indiscrimination, and exaggeration; the perverse mind	
indolent mind	怠心	rejection of the indolent mind is the foundation for doing the "acting"; includes violation, extravagance, slothfulness, and stealthiness; the idle mind	
indulgence (gratification)	嬌奢	one of the toxic factors, related to seduction	
infra-umbilical area	臍腹下	the site related to the large intestine; the anterior body part related to the lower-most burn-center	
inhale-gathering qi	吸聚之氣	The qi that promotes health for Taeyang type.	WHO-IST
insight	識見	the ability to adjust and cope with the given situation during changing circumstances	
inspection	行檢	the intelligence forged during the processes of self-control and regulation	
interior disease; interior-based disease	裏病	the pathology in which the Jeong damages the interior energy	
Interior-origin Small intestine disease (the same term in WHO-ICTM)	內觸小腸病	represented by regurgitation; the power to inspire (the cool Yin energy in the energy-fluid dimension) of the small intestine becomes weak, whereas the power to expire (the warm Yang energy in the energy-fluid dimension) of the esophagus becomes strong, and this causes the pathology in which the dispersing energy is excessive (in the dry esophagus) while the concentrating energy is unable to hold fort (in the middle burn-center), which is why the foodstuff is not readily accepted into the system and instead thrown up	WHO-IST

Term	Chinese	Definition / Description	Ref
jaw-chest-navel-abdomen	頷臆臍腹	the body part involved in doing the Knowing	
Jin-Go-Yu-AeK	津膏油液	the basic unit of biological activity derived from the warm-hot-cool-cold energy created in the Four Bu Organs; the basic composing unit of the Four Anterior Pools; the basic structural unit of the human body	
Kidney Affiliation	腎黨	comprises the large intestine, groin, mouth, pelvis, bone; the metabolic and circulatory pathway of the cold energy	
Kidney Heat-based Exterior Heat disease	腎受熱表熱病	a pathological condition caused by "largeness" of the kidney system; represented by the Ulgwang symptomatic pattern and the Mangyang symptomatic pattern, each of which can be divided into three phases (the initial, middle, and terminal stages)	WHO-IST
knack	材幹	the expressive motility related to actions that one becomes accustomed to physically	WHO-IST
knowledgeability (wisdom)	聞見	one of the tonic factors, avoidance of opulence	
large intestine	大腸	one of the Four Bu Organs, the site where the cold energy is formed when the heavier component of the cool energy descends	
large Kidney system and small Spleen system	腎大脾小	having a large Kidney system and a small Spleen system; the organ structural scheme found in the Soeum type	
large Liver system and small Lung system	肝大肺小	havingalargeLiversystemandasmallLungsystem;theorganstructuralscheme found in the Taeeum type	
large Lung system and small Liver system	肺大肝小	having a large Lung system and a small Liver system; the organ structural scheme found in the Taeyang type	
large Spleen system and small Kidney system	脾大腎小	having a large Spleen system and a small Kidney system; the organ structural scheme found in the Soyang type	
Liver Affiliation	肝黨	comprises the small intestine, navel, nose, lower back, and flesh; the metabolic and circulatory pathway of the cool energy	
Liver Heat-based Interior Heat disease	肝受熱裏熱病	pathological condition caused by "bigness" of the liver system and the	

Term	Chinese	Definition / Description	Ref
		resultant excessiveness in the inspirational and concentrative energy of the liver; can be divided into the Joyeol symptomatology and the Eumhyeol-mogal symptomatology	
loosening and releasing force	寬放之力	the driving force responsible for using the thick dregs of the pool of Hyeol; the loosening and releasing force of the waist is used to process the thick dregs of the pool of Hyeol and form flesh	WHO-IST
lower back and rib	腰脇	flank; a site pertinent to the liver	
lower back area	腰	the site related to the liver; the posterior body part related to the mid-lower burn-center	
lower burn-center; lowermost burn-center	下焦	one of the Four Burn-Centers, the sector related to the kidney and large intestine, extends to the concept of the Kidney Affiliation	
lower spine	腰脊	the site where the clear energy of the pool of Yu, after coming out by the driving force of the nose (force driven by smelling) to be thence converted into Hyeol, goes in to form the pool of Hyeol ; the body part pertinent to the mid-lower burn-center	
lower spine and hip	脊曲	lower spine; a site pertinent to the kidney	
Lung Affiliation	肺黨	comprises the esophagus, sublingua, ear, cranium, and skin; the metabolic and circulatory pathway of the warm energy	
maneuver	方略	the basic motility that includes both instinctive and adaptive movements	
manifestation of Jeong	情氣	the manifestation of Jeong in the form of energy in the four constitutional types	
manifestation of Seong	性氣	the manifestation of Seong in the form of energy in the four constitutional types	
measurement	度量	the inherent intelligence inseminated before birth	
mid-lower burn-center	中下焦	one of the Four Burn-Centers, the sector related to the liver and small intestine, extends to the concept of the Liver Affiliation	
mid-upper burn-center	中上焦	one of the Four Burn-Centers, the sector related to the spleen and	

Term	Chinese	Definition / Description	Ref
		stomach, extends to the concept of the Spleen Affiliation	
morphology-symptomatology	形證	a concept that incorporates the external morphology (the manifestations of the constitutionally differentiated pathology) and the internal physiopathology (the property and principle of internal organs causing the constitutionally differentiated physiology and the pathogenic factors of endogen, exogen, and psychological distress)	
muscle	筋	the end product generated when the thick dregs of the pool of Mak are processed by the driving force of the hand (the collecting and gathering force)	
nape area; supra-spinal area	顀下背上	the site related to the lung; the posterior body part related to the upper-most burn-center	
negligence (carelessness)	懶怠	one of the toxic factors, related to intoxication	
original symptomology	素病, 素證	the symptomatology observed commonly or usually in an individual	
palpitation	怔忡	the major pathology that can be seen when the cowardice of the Taeeumin escalates to greater cowardice, a moderate Taeeumin pathology	WHO-IST
pelvis (hip)	膀胱	the site where the clear energy of the pool of Aek, after coming out by the driving force of the mouth (force driven by tasting) to be thence converted into Jeong, goes in to form the pool of Jeong ; the body part pertinent to the lower burn-center	
pool of Aek; pool of thick fluid	液海	the pool formed when the cold energy converts into Aek in the large intestine and thereafter goes into the groin to form the anterior pool; serves as storage facility for Ji and is the root of the mouth	
pool of Go; pool of pasty fluid	膏海	the pool formed when the hot energy converts into Go in the stomach and thereafter goes into the intermammilla to form the anterior pool; serves as storage facility for Ryeo and is the root	

Term	Chinese	Definition / Description	Ref
		of the eye	
pool of Hyeol; pool of blood-like fluid	血海	the pool formed when the clear energy of the pool of Yu comes out through the nose and becomes Hyeol before going into the lower spine where it is condensed to form the posterior pool; serves as storage facility for Hon (魂) and is the root of the liver	
pool of Jeong; pool of essence-like fluid	精海	the pool formed when the clear energy of the pool of Aek comes out through the mouth and becomes Jeong before going into the pelvis where it is condensed to form the posterior pool; serves as storage facility for Baek (魄) and is the root of the kidney	
pool of Jin; pool of thin fluid	津海	the pool formed when the warm energy converts into Jin in the esophagus and thereafter goes into the sublingua to form the anterior pool; serves as storage facility for Ui and is the root of the ear	
pool of Mak; pool of membranous fluid	膜海	the pool formed when the clear energy of the pool of Go comes out through the eye and becomes Gi before going into the upper spine where it is condensed to form the posterior pool; serves as storage facility for Ryeong (靈) and is the root of the spleen	
pool of Ni; pool of greasy fluid	膩海	the pool formed when the clear energy of the pool of Jin comes out through the ear and becomes Sin before going into the cranium where it is condensed to form the posterior pool; serves as storage facility for Sin (神) and is the root of the lung	
pool of Yu; pool of oily fluid	油海	the pool formed when the cool energy converts into Yu in the small intestine and thereafter goes into the navel to form the anterior pool; serves as storage facility for Jo and is the root of the nose	
prejudice (partiality)	偏急	one of the toxic factors, related to authority	
principals of life preservation	保命之主	Refers to the main principals for the maintenance of health and enhancement of life expectancy for individual	

Term	Chinese	Definition / Description	Ref
		constitutions.	
proclivity and capacity; disposition and ability	性質材幹	a method of observing the mental attributes in order to discern the constitutional type of a given individual (along with the constant mindset and the predisposed greed); the proclivity (disposition) is the combined attribute of the inherent Seong and Jeong, and the capacity (ability) is the pattern of Humanly Affair (administration, association, aggregation, abidance) conducted through Jeong	WHO-IST
property and principle of internal organs	臟腑性理	the principles of expression and manifestation of Seong-Jeong that is also linked to the principles of organ formation ; the property and principle of internal organs; the "largeness" and "smallness" of the internal organs formed by constitutionally differentiated proclivity of Seong and Jeong	WHO-IST
raising-lowering-loosening-tightening	升降緩束	the therapeutic principles in treating constitutionally differentiated pathologies; the Soeumin and Soyangin pathologies are cold-hot energy based pathologies that require therapeutic approaches rooted in the superior-inferior dimension (or the "raising and lowering" approach), whereas the Taeeumin and Taeyangin pathologies are warm-cool energy based pathologies that require therapeutic approaches rooted in the interior-exterior dimension (or the "loosening and tightening" approach)	
ren-yi-li-zhi; four virtues	仁義禮智	the Four Virtues that are the antithesis of the Four Vices of ignobility, frivolity, avarice, and timidity; the virtues that can be discovered through the Four Indications	WHO-IST
respect (honor)	恭敬	the element connected to longevity in Lee Jema's paradigm of self-regulation	
rising momentum	起勢	the physical comportment reflective of the straightening and spreading property of the lung	
sagacious perception	慧覺	that which was given to man as an element of the Nature (性) when the	

Term	Chinese	Definition / Description	Ref
		universe was brought to life; the basis for projecting the Virtue; the specific method of doing the Knowing	
Sasang (Sixiang); the Four Images	四象	in cosmogony, the fundamental unit of structure and change in the universe based on the universality of the quaternity as the primary paradigm of cosmogony	
Sasang Constitutional Medicine	四象醫學	The branch of traditional Korean medicine initiated by Lee Je-ma, which stresses the theory of the four constitutions: physiology, pathology, diagnosis, and maintenance of health; also called Four Constitution Medicine.	
Sasangin; four constitutional types	四象人	the four constitutional types found in SCM: Taeyangin (Taeyang type), Soyangin (Soyang type), Taeeumin (Taeeum type), and Soeumin (Soeum type)	WHO-IST
sensory organs; ear-eye-nose-mouth	耳目鼻口	the physical organs responsible for perceiving the temporal, social, human, and natural environment pertinent to the Heavenly frame; the passageway for interacting with the Heavenly Loom elements; the receptor that perceives and receives external phenomena	WHO-IST
Seong-Jeong	性情	the differentiation and concretization of sorrow, anger, joy, and pleasure into Seong and Jeong; a concept established within the quadrifocal paradigm (quaternity)	
Seong-Jeong	性情	the differentiation and concretization of sorrow, anger, joy, and pleasure into Seong and Jeong; a concept established within the quadrifocal paradigm (quaternity)	
severity of disease	輕重險危	the level of severity of physiological conditions and pathological conditions; (becoming worse in the order of "light → mild → moderate → severe → critical")	
simplicity (austerity)	簡約	one of the tonic factors, avoidance of seduction	
Sin-Gi-Hyeol-Jeong	神氣血精	the basic unit of biological activity derived from the clear energy of the Four Anterior Pools that is pulled out	

Term	Chinese	Definition / Description	Ref
		to fill the Four Burn-Centers; the basic composing unit of the Four Posterior Pools; the basic functional unit of the human body	
Sin-Ryeong-Hon-Baek	神靈魂魄	the substance stored in the Four Posterior Pools, related to mental and psychological functioning	
sitting momentum	坐勢	the physical comportment reflective of the stacking and accumulating property of the kidney	
skin	皮毛	the end product generated when the thick dregs of the pool of Ni are processed by the driving force of the head (the straightening and stretching force)	
small intestine	小腸	one of the Four Bu Organs, the site where the cool energy is formed when the water-food coming down from the stomach is consumed and distilled	
Soeum type Congestive Hyperpsychotic initial phase pattern	少陰人 鬱狂初證	A pattern characterized by persisting systemic heat, cold-intolerance, absence of perspiration, and personality changes to more fretful or restless mentality in the Soeum type Congestive Hyperpsychotic pattern. It may be explained by: The Yang energy of the Kidney sector (Lower sector) fails to ascend and to connect to the Spleen sector (Mid-upper sector) due to pathogenic cold, and then it begins to be confined in the Infra-spinal sector (dorsal Lower sector). There is no damage to the Yang energy of the Spleen sector (Mid-upper sector).	WHO-ICTM
Soeum type Congestive Hyperpsychotic intermediate phase pattern	少陰人 鬱狂中證	A pattern characterized by systemic heat, absence of cold-intolerance, and absence of perspiration that are superimposed on congested-stomach related symptoms of constipation and vexation in the Soeum type Congestive Hyperpsychotic pattern. It may be explained by: The Yang energy of the Kidney sector (Lower sector) fails to ascend and to connect to the Spleen sector (Mid-upper sector) due to severe pathogenic cold of the Large Intestine	WHO-ICTM

Term	Chinese	Definition / Description	Ref
		sector (Frontal Lower sector), and then it is confined in the Infra-spinal sector (dorsal Lower sector). Also strengthening pathogenic cold of the Large Intestine sector (Frontal Lower sector) disturbs the warm energy of the Stomach sector (Frontal Mid-upper sector), but does not damage the Yang energy of the Spleen sector (Mid-upper sector).	
Soeum type Congestive Hyperpsychotic pattern	少陰人 鬱狂證	A pattern characterized by systemic heat, somatic pain (Including headache), and absence of perspiration. It may be explained by: The Yang energy of the Kidney sector (Lower sector) fails to ascend and to connect to the Spleen sector (Mid-upper sector) due to pathogenic cold, and then it is confined in the Infra-spinal sector (dorsal Lower sector). There is no damage to the Yang energy of the Spleen sector (Mid-upper sector).	WHO-ICTM
Soeum type Congestive Hyperpsychotic pattern, NEC	少陰人 鬱狂證, 他處未分類	A pattern characterized by systemic heat, somatic pain (Including headache), and absence of perspiration. It may be explained by the subpattern which is not else classified in the Soeum type Congestive Hyperpsychotic pattern.	WHO-ICTM
Soeum type Congestive Hyperpsychotic terminal phase pattern	少陰人 鬱狂末證	A pattern characterized by systemic heat., absence of cold-intolerance, absence of perspiration, and constipated bowel passage that are superimposed on symptoms of tidal heat, delirious muttering, panting, and rolled-up eyes in the Soeum type Congestive Hyperpsychotic pattern. It may be explained by: The Yang energy of the Kidney sector (Lower sector) completely fails to ascend and to connect to the Spleen sector (Mid-upper sector) due to intense pathogenic cold and it is confined in the Infra-spinal sector (dorsal Lower sector). The pathogenic cold of the Large Intestine sector (Frontal Lower sector) also intensifies, but does not damage the Yang energy	WHO-ICTM

Term	Chinese	Definition / Description	Ref
		of the Spleen sector (Mid-upper sector).	
Soeum type Exterior Interior combined pattern	少陰人 表裏兼病證	A pattern characterized by exterior symptoms that include systemic heat, headache, somatic pain; interior symptoms that include abdominal pain and diarrhea (bowel irritability) at a same time. It may be explained by: The Yang energy of the Kidney sector (Lower sector) fails to ascend and to connect to the Spleen sector (Mid-upper sector), concurently the Warm energy of the Stomach sector (Frontal Mid-upper sector) is damaged by pathogenic cold.	WHO-ICTM
Soeum type Exterior Interior combined pattern with exterior dominance	少陰人 表病兼裏證	A pattern characterized by exterior symptoms that include systemic heat, headache, somatic pain; interior symptoms that include abdominal pain and diarrhea (bowel irritability) secondarily. It may be explained primarily as pathogenic cold causes the Yang energy of the Kidney sector (Lower sector) to fail to ascend and to connect to the Spleen sector (Mid-uppersector). Secondarily as pathogenic cold damages the Warm energy of the Stomach sector (Frontal Mid-upper sector).	WHO-ICTM
Soeum type Exterior Interior combined pattern with interior dominance	少陰人 裏病兼表證	A pattern characterized by interior symptoms that include abdominal pain and diarrhea (bowel irritability), etc; exterior symptoms that include systemic heat, headache, somatic pain secondarily. It may be explained primarily as pathogenic cold damages the Warm energy of the Stomach sector (Frontal Mid-upper sector). Secondarily as pathogenic cold causes the Yang energy of the Kidney sector (Lower sector) to fail to ascend and to connect to the Spleen sector (Mid-upper sector).	WHO-ICTM
Soeum type Greater Yang disease Reverting Yin pattern	少陰人 太陽病 厥陰證	A pattern characterized by coldness of extremities, Lower abdominal pain & fullness, curled tongue and retracted testicles, systemic heat, cold-intolerance, presence of perspiration. It may be	WHO-ICTM

Term	Chinese	Definition / Description	Ref
		explained by: Persistent Initial Phase Congestive Hyperpsychotic pattern or persistent Initial Phase Yang Depletion pattern (exhibiting symptoms of cold-intolerance, systemic heat, and hyperhidrosis) that is prolonged beyond usual term and is not resolved.	
Soeum type Greater Yin pattern	少陰人 太陰證	A pattern characterized by abdominal pain and diarrhea, and converse absence of dry mouth, somatic pain, and agitation. It may be explained by: Damage of the Warm energy of the Stomach sector (Frontal Mid-upper sector) that is oppressed by the pathogenic cold of the Large Intestine sector (Frontal Lower sector), though not threatening the Yang energy of the Spleen sector (Mid-upper sector) itself.	WHO-ICTM
Soeum type Greater Yin pattern accompanying abdominal pain and bowel irritability	少陰人 太陰病 腹痛自利證	A pattern characterized by impaired digestion, exacerbated diarrhea (bowel irritability), accompanied by abdominal pain, but without dry mouth in the Soeum type Greater Yin pattern. It may be explained as a pattern of the Large Intestine sector (Frontal Lower sector) when intensifying pathogenic cold of the Large Intestine sector (Frontal Lower sector) oppresses and damages the Warm energy of the Stomach sector (Frontal Mid-upper sector).	WHO-ICTM
Soeum type Greater Yin pattern accompanying Edema	少陰人 太陰病 浮腫證	A pattern characterized by edema in the Soeum type Greater Yin pattern. It may be explained as a pattern of the Stomach sector (Frontal Mid-upper sector) when intense pathogenic cold of the Large Intestine sector (Frontal Lower sector) oppresses and damages the Warm energy of the Stomach sector (Frontal Mid-upper sector). The body's ability to metabolize fluid decreases because of Stomach sector (Frontal Mid-upper sector) dysfunction.	WHO-ICTM
Soeum type Greater Yin pattern accompanying epigastric block and	少陰人 太陰病 乾藿亂關格病證	A pattern characterized by rapidly occuring symptoms of abdominal pain, abdominal distention, and cold extremities, without accompanying	WHO-ICTM

Term	Chinese	Definition / Description	Ref
repulsion disease		vomiting or diarrhea in the Soeum type Greater Yin pattern.It may be explained as a pattern of the Stomach sector (Frontal Mid-upper sector) when intense pathogenic cold of the Large Intestine sector (Frontal Lower sector) rapidly damages the Warm energy of the Stomach sector (Frontal Mid-upper sector).	
Soeum type Greater Yin pattern accompanying Epigastric stuffiness and fullness	少陰人 太陰病 心下痞滿證	A pattern characterized by stuffiness and fullness in the chest or epigastric area, without signs of epigastric hardening and binding in the Soeum type Greater Yin pattern. It may be explained as a pattern of the Stomach sector (Frontal Mid-upper sector) when pathogenic cold of the Large Intestine sector (Frontal Lower sector) oppresses the Warm energy of the Stomach sector (Frontal Mid-upper sector) causing Stomach sector (Frontal Mid-upper sector) disorders.	WHO-ICTM
Soeum type Greater Yin pattern accompanying Jaundice	少陰人 太陰病 黃疸證	A pattern characterized by jaundice in the Soeum type Greater Yin pattern. It may be explained as a pattern of the Stomach sector (Frontal Mid-upper sector) when intense pathogenic cold of the Large Intestine sector (Frontal Lower sector) oppresses and damages the Warm energy of the Stomach sector (Frontal Mid-upper sector) which allows the diffusion of pathogens throughout the whole body because of Stomach sector (Frontal Mid-upper sector) dysfunction.	WHO-ICTM
Soeum type Greater Yin pattern accompanying Visceral bind	少陰人 太陰病 藏結證	A pattern characterized by the unfavorable prognostic sign which is hardening and binding in the epigastric area in the Soeum type Greater Yin pattern. It may be explained as a pattern of the Stomach sector (Frontal Mid-upper sector) when intense pathogenic cold of the Large Intestine sector (Frontal Lower sector) persistently oppresses and damages the Warm energy of the Stomach sector	WHO-ICTM

Term	Chinese	Definition / Description	Ref
		(Frontal Mid-upper sector), and develops into a chronic condition.	
Soeum type Greater Yin pattern by Yintoxin	少陰人 太陰病 陰毒證	A pattern characterized by somatic pain that is similar to being thrashed with club, bluish complexion, and coldness of extremities in the Soeum type Greater Yin pattern. It may be explained by: Persistent Initial Phase Greater Yin pattern that is prolonged beyond usual term and not resolved.	WHO-ICTM
Soeum type Greater Yin pattern, NEC	少陰人 太陰證, 他處未分類	A pattern characterized by abdominal pain and diarrhea, but absence of dry mouth, somatic pain & agitation. It may be explained by the subpattern which is not else classified in the Soeum type Greater Yin pattern	WHO-ICTM
Soeum type Lesser Yin pattern	少陰人 少陰證	A pattern characterized by abdominal pain and diarrhea, and presence of dry mouth, headache, somatic pain, and agitation. It may be explained by: Damage of the Yang energy of the Spleen sector (Mid-upper sector) as well as the Warm energy of the Stomach sector (Frontal Mid-upper sector) due to intense pathogenic cold of the Large Intestine sector (Frontal Lower sector).	WHO-ICTM
Soeum type Lesser Yin pattern accompanying abdominal pain and bowel irritability	少陰人 少陰病 腹痛自利證	A pattern characterized by abdominal pain and diarrhea accompanied by dry mouth, headache, somatic pain, and agitation in the Soeum type Lesser Yin pattern. It may be explained as a pattern of the Large Intestine sector (Frontal Lower sector) when intense pathogenic cold of Large Intestine sector (Frontal Lower sector) oppress and damage the Yang energy of the Spleen sector (Mid-upper sector) as well as the Warm energy of the Stomach sector (Frontal Mid-upper sector).	WHO-ICTM
Soeum type Lesser Yin pattern accompanying green tinged watery diarrhea	少陰人 少陰病 下利青水證	A pattern characterized by green-tinged, watery diarrhea and dry mouth in the Soeum type Lesser Yin pattern. It may be explained as a pattern characterized by watery diarrhea due to intense pathogenic cold of the Large Intestine sector (Frontal Lower sector), while the	WHO-ICTM

Term	Chinese	Definition / Description	Ref
		Yang energy of the Spleen sector (Mid-upper sector) as well as the Warm energy of Stomach sector (Frontal Mid-upper sector) are damaged and oppressed by the pathogenic cold of Large Intestine sector (Frontal Lower sector).	
Soeum type Lesser Yin pattern accompanying Visceral collapse	少陰人 少陰病 藏厥證	A pattern characterized by severe coldness of extremities and stuffiness that leads to fretfulness and distraction in the Soeum type Lesser Yin pattern. It may be explained by: Severe damage of the Yang energy of the Spleen sector (Mid-upper sector) and of the Warm energy of Stomach sector (Frontal Lower sector). Inability to overcome the damaging effects of the pathogenic cold of Large Intestine sector (Frontal Lower sector).	WHO-ICTM
Soeum type Lesser Yin pattern by Exuberant Yin and repelling Yang	少陰人 少陰病 陰盛隔陽證	A pattern characterized by severe coldness of extremities and converse flushing in the face and head combined with uncontainable feeling of stuffiness. in the Soeum type Lesser Yin pattern. It may be explained by: Severe damage of the Yang energy of the Spleen sector (Mid-upper sector) as well as the Warm energy of the Stomach sector (Frontal Mid-upper sector). The pathogenic cold of the Large Intestine sector (Frontal Lower sector) increases and strengthens.	WHO-ICTM
Soeum type Lesser Yin pattern, NEC	少陰人 少陰證, 他處未分類	A pattern characterized by abdominal pain and diarrhea accompanied by the dry mouth, headache, somatic pain, and agitation. It may be explained by the subpattern which is not else classified in the Soeum type Lesser Yin pattern.	WHO-ICTM
Soeum type Yang Depletion initial phase pattern	少陰人 亡陽初證	A pattern characterized by systemic heat, cold-intolerance, and perspiration in the Soeum type Yang Depletion pattern. It may be explained by: The Yang energy of the Kidney sector (Lower sector) fails to ascend and to connect to the Spleen sector (Mid-upper sector) due to pathogenic cold, and then it begins to become confined in the Infra-spinal sector (dorsal Lower sector).	WHO-ICTM

Term	Chinese	Definition / Description	Ref
		The Yang energy of the Kidney sector (Lower sector) damages the Yang energy of the Spleen sector (Mid-upper sector).	
Soeum type Yang Depletion intermediate phase pattern	少陰人 亡陽中證	A pattern characterized by systemic heat, constipated bowel passage that are superimposed on absence of cold-intolerance, presence of heat-intolerance, and perspiration in the Soeum type Yang Depletion pattern. It may be explained by: The Yang energy of the Kidney sector (Lower sector) fails to ascend and to connect to the Spleen sector (Mid-upper sector) due to severe pathogenic cold of the Large Intestine sector (Frontal Lower sector), and then it is confined in the Infra-spinal sector (dorsal Lower sector). The strengthening pathogenic cold of the Large Intestine sector (Frontal Lower sector) disturbs the warm energy of the Stomach sector (Frontal Mid-upper sector), further damaging the Yang energy of the Spleen sector (Mid-upper sector).	WHO-ICTM
Soeum type Yang Depletion pattern	少陰人 亡陽證	A pattern characterized by systemic heat, somatic pain (Including headache), and presence of perspiration. It may be explained by: The Yang energy of the Kidney sector (Lower sector) fails to ascend and to connect to the Spleen sector (Mid-upper sector) due to pathogenic cold,and then it is confined in the Infra-spinal sector (dorsal Lower sector. The Yang energy of the Kidney sector (Lower sector) damages the Yang energy of the Spleen sector (Mid-upper sector).	WHO-ICTM
Soeum type Yang Depletion pattern NEC	少陰人 亡陽證, 他處未分類	A pattern characterized by systemic heat, somatic pain (Including headache), and presence of perspiration. It may be explained by the subpattern which is not else classified in the Soeum type Yang Depletion pattern.	WHO-ICTM
Soeum type Yang Depletion terminal phase pattern	少陰人 亡陽末證	A pattern characterized by systemic heat, cold-intolerance, and constipated bowel passage that are superimposed on the symptom of profuse perspiration in	WHO-ICTM

Term	Chinese	Definition / Description	Ref
		the Soeum type Yang Depletion pattern. It may be explained by: The Yang energy of the Kidney sector (Lower sector) completely fails to ascend and to connect to the Spleen sector (Mid-upper sector) due to intense pathogenic cold and it is confined in the Infra-spinal sector (dorsal Lower sector). The acutely strengthening pathogenic cold of the Large Intestine sector (Frontal Lower sector) damages the Yang energy of the Spleen sector (Mid-upper sector).	
Soeumin; Soeum type; SE type	少陰人	one who is given the organ structural scheme of a large Kidney system and a small Spleen system	WHO-IST
somatic morphology and physical comportment	體形氣像	a passive method of observing the physical attributes in order to discern the constitutional type of a given individual; the differences in external appearances that arise from the differences in the internal organ scheme due to the deviation in the Seong-Jeong of sorrow-anger-joy-pleasure	WHO-IST
somatic organs; lung-spleen-liver-kidney	肺脾肝腎	the physical organs responsible for expressing the Humanly Affair; the passageway for interacting with the Humanly Affair elements; the vehicle that conducts living	
Soyang type Chest Heat congested pattern	少陽人 胸隔熱證	A pattern characterized by systemic heat, constipated bowel movement, and thirst. It may be explained by: The Clear-Yang of the Large Intestine sector (Frontal Lower sector) fails to ascend to the Stomach sector (Frontal Mid-upper sector), extremities and head specifically the face. Intense pathogenic heat is formed in the Stomach sector (Frontal Mid-upper sector).	WHO-ICTM
Soyang type Chest Heat congested pattern, NEC	少陽人 胸隔熱證, 他處未分類	A pattern characterized by systemic heat, constipated bowel movement, and thirst. It may be explained by the subpattern which is not else classified in the Soyang type Chest Heat congested pattern.	WHO-ICTM
Soyang type Chest-binding pattern	少陽人 結胸證	A pattern characterized by cold-intolerance, nausea and retching,	WHO-ICTM

Term	Chinese	Definition / Description	Ref
		vomiting when drinking water, shortness of breath, dyspnea, fullness in the chest and hypochondrium, dry thirst and delirious speech, headache, perspiration in head, numbness and coldness in joints in the Soyang type Lesser Yang Wind Damage pattern. It may be explained by: The chronic state of the Yin energy of the Spleen sector (Mid-upper sector) failing to descend to the Kidney sector (Lower sector) due to intensive pathogenic heat which results in Yin energy confinement in the upper back area of the dorsal Mid-upper sector. The pathogenic heat of the Stomach sector (Frontalal Mid-upper sector) is strongly strengthened.	
Soyang type Exterior Interior combined pattern	少陽人 表裏兼病證	A pattern characterized by exterior symptoms that include cold-intolerance, diarrhea, somatic pain, dizziness, headache, etc.; interior symptoms that include heat intolerance, constipated bowel movement at a same time. It may be explained by: The Yin energy of the Spleen sector (Mid-upper sector) fails to descend and to connect to the Kidney sector (Lower sector), concurently the Clear Yang (Clear ascending Yang energy) of the Large Intestine sector (Frontal Lower sector) is damaged by pathogenic heat.	WHO-ICTM
Soyang type Exterior Interior combined pattern with exterior dominance	少陽人 表病兼裏證	A pattern characterized by exterior symptoms that include cold-intolerance, diarrhea, somatic pain, dizziness, headache, etc.; interior symptoms that include heat intolerance, constipated bowel movement secondarily. It may be explained primarily as pathogenic heat causes the Yin energy of the Spleen sector (Mid-upper sector) to fail to descend and to connect to the Kidney sector (Lower sector), secondarily as pathogenic heat damages the Clear Yang (Clear ascending Yang energy) of the Large Intestine sector (Frontal Lower sector).	WHO-ICTM

Term	Chinese	Definition / Description	Ref
Soyang type Exterior Interior combined pattern with interior dominance	少陽人 裏病兼表證	A pattern characterized by interior symptoms that include heat intolerance, constipated bowel movement, etc; exterior symptoms that include cold-intolerance, diarrhea, somatic pain, dizziness, headache secondarily. It may be explained primarily as pathogenic heat damages the Clear Yang (Clear ascending Yang energy) of the Large Intestine sector (Frontal Lower sector), secondarily as pathogenic heat causes the Yin energy of the Spleen sector (Mid-upper sector) to fail to descend and to connect to the Kidney sector (Lower sector).	WHO-ICTM
Soyang type Lesser Yang Wind Damage mild pattern	少陽人 少陽傷風輕證	A pattern characterized by bitter taste in mouth, dry throat, dizziness, deafness, retching, and alternating chills and fever in the Soyang type Lesser Yang Wind Damage pattern. It may be explained by: The inital stages of the Yin energy of the Spleen sector (Mid-upper sector) failing to descend to the Kidney sector (Lower sector) due to pathogenic heat which results in Yin energy confinement in the Upper Back area (dorsal Mid-upper sector).	WHO-ICTM
Soyang type Lesser Yang Wind Damage pattern	少陽人 少陽傷風證	A pattern characterized by systemic heat, cold-intolerance, somatic pain, vexation, bitter taste in mouth, dry throat, dizziness, headache, deafness. It may be explained by: The failure of the Yin energy of the Spleen sector (Mid-upper sector) to descend to the Kidney sector (Lower sector) due to pathogenic heat which results in Yin energy confinement to the Upper Back area (dorsal Mid-upper sector). There is no damage to the Yin energy of the Kidney sector (Lower sector).	WHO-ICTM
Soyang type Lesser Yang Wind Damage pattern, NEC	少陽人 少陽傷風證, 他處未分類	A pattern characterized by systemic heat, cold-intolerance, somatic pain, vexation, bitter taste in mouth, dry throat, dizziness, headache, deafness. It may be explained by the subpattern which is not else classified in the	WHO-ICTM

Term	Chinese	Definition / Description	Ref
		Soyang type Lesser Yang Wind Damage pattern	
Soyang type Lower Wasting Thirst pattern	少陽人 下消證	A pattern characterized by thirst, agitation, emaciation of the legs, turbid urine in the Soyang type Yin Deficit pattern. It may be explained by: The Clear-Yang of the Large Intestine sector (Frontal Lower sector) is weak. The initial stages of the weakening of the Yin energy of the Kidney sector (Lower sector).	WHO-ICTM
Soyang type Middle Wasting Thirst pattern	少陽人 中消證	A pattern characterized by hyperactive digestive functions and increased appetite in the Soyang type Chest Heat congested pattern. It may be explained by: Pathogenic heat forms in the Stomach sector (Frontal Mid-upper sector). The Clear-Yang of the Large Intestine sector is weak; therefore, it fails to reach the Stomach sector (Frontal Mid-upper sector).	WHO-ICTM
Soyang type Stomach Heat pattern	少陽人 胃熱證	A pattern characterized by intense Heat-related symptoms accompanied by constipated bowel movement in the Soyang type Chest Heat congested pattern. It may be explained by: The Clear-Yang of the Large Intestine sector (Frontal Lower sector) successfully ascends to the Stomach sector (Frontal Mid-upper sector). Pathogenic heat in the Stomach sector (Frontal Mid-upper sector) increases in severity and is excessive.	WHO-ICTM
Soyang type Upper Wasting Thirst pattern	少陽人 上消證	A pattern characterized by dry mouth, red tongue, increased fluid ingestion (polydipsia), and more frequent and concentrated urine in the Soyang type Chest Heat congested pattern. It may be explained by: Pathogenic heat forms in the Stomach sector (Frontal Mid-upper sector). The Clear-Yang of the Large Intestine sector is slightly weakened; therefore, it can ascend to the Stomach sector, but fails to reach the extremities and the head, specifically the face.	WHO-ICTM
Soyang type Yin	少陽人 陰虛午熱證	A pattern characterized by diurnal	WHO-

Term	Chinese	Definition / Description	Ref
Deficit Diurnal Heat pattern		flaring-up of systemic heat, thirst, coldness in the dorsal region, and retching in the Soyang type Yin Deficit pattern. It may be explained by: The Clear-Yang of the Large Intestine sector (Frontal Lower sector) is weak. A chronic state of weakness with severe damage to the Yin energy of the Kidney sector (Lower sector).	ICTM
Soyang type Yin Deficit pattern	少陽人 陰虛證	A pattern characterized by diurnal flaring-up of systemic heat, thirst, coldness in the dorsal region, and retching, emaciation of the legs, turbid urine. It may be explained by: The Clear-Yang of the Large Intestine sector (Frontal Lower sector) is weak eventually leading to a weakening of the Yin energy of the Kidney sector (Lower sector).	WHO-ICTM
Soyang type Yin Deficit pattern, NEC	少陽人 陰虛證, 他處未分類	A pattern characterized by diurnal flaring-up of systemic heat, thirst, coldness in the dorsal region, and retching, emaciation of the legs, turbid urine. It may be explained by the subpattern which is not else classified in the Soyang type Yin Deficit pattern.	WHO-ICTM
Soyang type Yin Depletion pattern	少陽人 亡陰證	A pattern characterized by diarrhea, often accompanied by either heat-related patterns with headache or cold-related patterns with stomachache. It may be explained by: The failure of the Yin energy of the Spleen sector (Mid-upper sector) to descend to the Kidney sector (Lower sector) due to pathogenic heat which eventually leads to damage of the Yin energy of the Kidney sector (Lower sector).	WHO-ICTM
Soyang type Yin Depletion pattern accompanYing Cold related Stomachache	少陽人 身寒腹痛 亡陰證	A pattern characterized by cold-related patterns, stomachache, and diarrhea in the Soyang type Yin Depletion pattern. It may be explained by: The failure of the Yin energy of the Spleen sector (Mid-upper sector) to descend to the Kidney sector (Lower sector) due to pathogenic heat which is not severe. Though pathogenic heat is not intense,	WHO-ICTM

Term	Chinese	Definition / Description	Ref
		Yin energy of the Spleen sector (Mid-upper sector) is very weak and eventually leads to severe damage of the Yin energy of the Kidney sector (Lower sector).	
Soyang type Yin Depletion pattern accompanYing Heat related Headache	少陽人 身熱頭痛亡陰證	A pattern characterized by heat-related patterns, headache, and diarrhea in the Soyang type Yin Depletion pattern. It may be explained by: The failure of the Yin energy of the Spleen sector (Mid-upper sector) to descend to the Kidney sector (Lower sector) due to an increase in severity of pathogenic heat which eventually leads to damage of the Yin energy of the Kidney sector (Lower sector).	WHO-ICTM
Soyang type Yin Depletion pattern, NEC	少陽人 亡陰證, 他處未分類	A pattern characterized by diarrhea, often accompanied by either heat-related patterns with headache or cold-related patterns with stomachache. It may be explained by the subpattern which is not else classified in the Soyang type Yin Depletion pattern	WHO-ICTM
Soyangin; Soyang type; SY type	少陽人	one who is given the organ structural scheme of a large Spleen system and a small Kidney system	WHO-IST
Spleen Affiliation	脾黨	comprises the stomach, intermammilla, eye, upper back, and muscle; the metabolic and circulatory pathway of the hot energy	
Spleen Cold-based Exterior Cold disease	脾受寒表寒病	pathological condition due to oppression of the exterior Yin and its consequent failure to descend, is distinguished into the Soyang-sangpung symptomatology and the Mangeum symptomatology	WHO-IST
standing momentum	立勢	the physical comportment reflective of the tolerating and slackening property of the liver	
state of (or indication of) healthiness; physiological condition	完實無病	the physiological conditions of each constitutional type; healthy, disease-free state with sufficient exterior and interior energy; physiological condition with a satisfactory level of the requisite energy	WHO-IST
stomach	胃	one of the Four Bu Organs, the site where the hot energy is formed when	

Term	Chinese	Definition / Description	Ref
		the water-food coming down from the esophagus is contained and steamed	
Stomach Cold-based Interior Cold disease	胃受寒裏寒病	a pathological condition caused by "smallness" of the spleen system; represented by the Taeeum symptomatic pattern and the Soeum symptomatic pattern that share the common symptoms of abdominal pain and bowel irritability	WHO-IST
Stomach Heat-based Interior Heat disease	胃受熱裏熱病	pathological condition due to failure of the clear Yang to rise harmoniously which then becomes overheated and generates exhausted Yang (scorching energy), is distinguished into the Hyunggyeok-yeol symptomatology and the Eumheo-oyeol symptomatology	WHO-IST
straightening and stretching force	直伸之力	the driving force responsible for using the thick dregs of the pool of Ni; the straightening and stretching force of the head is used to process the thick dregs of the pool of Ni and form skin	
strategy	籌策	the intelligence capable of planning and operating adequately and purposefully with a specific aim in view	
supra-thoracic area	頷下胸上	the site related to the esophagus; the anterior body part related to the upper-most burn-center	
Taeeum type Dryness Heat pattern	太陰人 燥熱證	A pattern characterized by black discoloration and parching of the fingertips and nummular erythematous patches and sores over the whole body, Constipation, frequent urination, thirst, consumptive overexertion, tinnitus, dimvision, weakened legs, and Lower back pain. It may be explained by: An excessiveness in the inspirational concentrative energy of the Liver sector (Mid-lower sector), called the Liver-heat state, in addition to the weakening of the expirational dispersive energy of the Lung sector (Upper sector), called the Lung-dryness state, producing an imbalance of energy and fluid metabolism, followed by the abnormal comsumption of energy and fluid.	WHO-ICTM
Taeeum type Dryness	太陰人 裏病	A pattern characterized by consumptive	WHO-

Term	Chinese	Definition / Description	Ref
Heat pattern accompanying Consumptive overexertion and dream emission	虛勞夢泄證	overexertion and accompanying dream emission in the Taeeum type Dryness Heat pattern. It may be explained by: A severely decreased capbility of the Lung sector (Upper sector) to expire and to disperse energy and fluid, and an excessiveness in the inspirational concentrative energy of the Liver sector (Mid-lower sector), followed by stagnation and the abnormal consumption of energy and fluid in the whole body.	ICTM
Taeeum type Dryness Heat pattern accompanying Excessive fluid discharge	太陰人 飮一溲二證	A pattern characterized by relatively increased urine output compared to the fluid intake, often accompanied by drier, waterless change in stool texture in the Taeeum type Dryness Heat pattern. It may be explained by: A malfunction of fluid metabolism, followed by the abnormal comsumption of fluid, with an excessiveness in the inspirational concentrative energy of the Liver sector (Mid-lower sector), and with the weakening of the expirational dispersive energy of the Lung sector (Upper sector).	WHO-ICTM
Taeeum type Dryness Heat pattern accompanying Finger blackening parching Erythematous eruption	太陰人 手指焦黑 癍瘡病證	A pattern characterized by black discoloration and parching of the fingertips and nummular erythematous patches and sores over the whole body in the Taeeum type Dryness Heat pattern. It may be explained by: An imbalance of energy and fluid metabolism, followed by the abnormal comsumption of energy and fluid in the skin and flesh throughout the whole body skin and flesh, with an excessiveness in the inspirational concentrative energy of the Liver sector (Mid-lower sector) and with the weakening of the expirational dispersive energy of the Lung sector (Upper sector).	WHO-ICTM
Taeeum type Dryness Heat pattern with Yin Blood Consumption	太陰人 陰血耗竭證	A pattern characterized by tinnitus, dim vision, weakening of legs, and Lower back pain in the Taeeum type Dryness Heat pattern. It may be explained by:	WHO-ICTM

Term	Chinese	Definition / Description	Ref
symptoms		An abnormal consumption of energy and fluid and sensory organ and musculoskeletal system disorders with an excessiveness in the inspirational concentrative energy of the Liver sector (Mid-lower sector), and with the weakening of the expirational dispersive energy of the Lung sector (Upper sector).	
Taeeum type Dryness Heat pattern, NEC	太陰人 燥熱證, 他處未分類	A pattern characterized by black discoloration and parching of the fingertips and nummular erythematous patches and sores over the whole body, Constipation, frequent urination, thirst, consumptive overexertion, tinnitus, dim vision, weakened legs, and Lower back pain. It may be explained by the subpattern which is not else classified in the Taeeum type Dryness Heat pattern.	WHO-ICTM
Taeeum type Esophagus Cold pattern	太陰人 胃脘寒證	A pattern characterized by post-prandial stuffiness and fullness and leg weakness, coughing, edema etc, which is based on symptoms of diarrhea, absence of perspiration, palpitation, dyspnea (labored breathing), plum-pit, pale complexion. It may be explained by: A weakening of the expirational dispersive energy of the Lung sector (Upper sector), in addition to a decreased capbility of the Esophagus sector (Frontal Upper sector) to expire and to disperse energy and fluid.	WHO-ICTM
Taeeum type Esophagus Cold pattern accompanying Abdominal distention and Edema	太陰人 腹脹浮腫病證	A pattern characterized by abdominal distention and edema in the Taeeum type Esophagus Cold pattern. It may be explained by: Stagnation of energy and fluid in the whole body including the peritonial cavity with a weakening in the expirational dispersive energy of the Lung sector (Upper sector) and Esophagus sector (Frontal Upper sector).	WHO-ICTM
Taeeum type Esophagus Cold pattern accompanying Chest and abdominal pain	太陰人 胸腹痛病證	A pattern characterized by chest pain and abdominal pain in the Taeeum type Esophagus Cold pattern. It may be explained by: Stagnation of energy and fluid in the thorax and the abdominal region with a weakening in the	WHO-ICTM

Term	Chinese	Definition / Description	Ref
		expirational dispersive energy of the Lung sector (Upper sector) and Esophagus sector (Frontal Upper sector).	
Taeeum type Esophagus Cold pattern accompanying Consumptive overexertion and dream emission	太陰人 表病 虛勞夢泄證	A pattern characterized by consumptive disease and accompanying dream emission in the Taeeum type Esophagus Cold pattern. It may be explained by: A severely decreased capbility of the Esophagus sector (Frontal Upper sector) and Lung sector (Upper sector) to expire and to disperse energy and fluid, followed by the the stagnation of and abnormal consumption of energy and fluid in the whole body.	WHO-ICTM
Taeeum type Esophagus Cold pattern accompanying Cough	太陰人 咳嗽病證	A pattern characterized by coughing in the Taeeum type Esophagus Cold pattern. It may be explained by: An imbalance in the respiratory function (decreased respiratory function), caused by the stagnation of energy and fluid with a weakening in the expirational dispersive energy of the Lung sector (Upper sector) and Esophagus sector (Frontal Upper sector).	WHO-ICTM
Taeeum type Esophagus Cold pattern accompanying Diarrhea	太陰人 泄瀉病證	A pattern characterized by diarrhea in the Taeeum type Esophagus Cold pattern. It may be explained by: A decrease in the function of energy-fluid metabolism in the Small Intestines with a weakening in the expirational dispersive energy of the Lung sector (Upper sector) and Esophagus sector (Frontal Upper sector).	WHO-ICTM
Taeeum type Esophagus Cold pattern accompanying Pestilence	太陰人 胃脘寒證 瘟病	A pattern characterized by cold-intolerance, systemic heat, somatic pain, and laryngopharyngeal swelling and pain in the Taeeum type Esophagus Cold pattern. It may be explained by: When signs and symptoms of plague-like infectious disease are present with a weakening in the expirational dispersive energy of the Lung sector (Upper sector) and Esophagus sector (Frontal Upper sector).	WHO-ICTM
Taeeum type Esophagus Cold	太陰人 食後痞滿腿脚無力	A pattern characterized by post-prandial stuffiness and fullness and leg weakness	WHO-ICTM

Term	Chinese	Definition / Description	Ref
pattern accompanying Postprandial stuffiness and fullness and Leg weakness	病證	in the Taeeum type Esophagus Cold pattern. It may be explained by: When signs and symptoms of the stagnation of fluid in the lower limbs and the malfunction of energy metabolism in the digestive system with a weakening in the expirational dispersive energy of the Lung sector (Upper sector) and Esophagus sector (Frontal Upper sector).	
Taeeum type Esophagus Cold pattern, NEC	太陰人 胃脘寒證, 他處未分類	A pattern characterized by post-prandial stuffiness and fullness and leg weakness, coughing, edema etc, which is based on symptoms of diarrhea, absence of perspiration, palpitation, dyspnea (labored breathing), plum-pit, pale complexion. It may be explained by the subpattern which is not else classified in the Taeeum type Esophagus Cold pattern.	WHO-ICTM
Taeeum type Exterior Interior combined pattern	太陰人 表裏兼病證	A pattern characterized by exterior symptoms that include cold-intolerance (accompanied with the absence of perspiration), dyspnea (labored breathing), somatic pain, diarrhea; interior symptoms that include systemic heat, ocular pain, dryness of nose, sleep disorder, tidal sweating, constipated bowel movement, thirst at a same time. It may be explained by: the damage of the Esophagus sector (Frontal Upper sector) which expires and disperses energy and fluid, and concurently an excessive condensation and concentration of the energy-fluid of the Liver sector (Mid-lower sector).	WHO-ICTM
Taeeum type Exterior Interior combined pattern with exterior dominance	太陰人 表病兼裏證	A pattern characterized by exterior symptoms that include cold-intolerance (accompanied with the absence of perspiration), dyspnea (labored breathing), somatic pain, diarrhea, etc ; interior symptoms that include systemic heat, ocular pain, dryness of nose, sleep disorder, tidal sweating, constipated bowel movement, thirst secondarily. It may be explained by: primarily the damage of the Esophagus sector (Frontal	WHO-ICTM

Term	Chinese	Definition / Description	Ref
		Upper sector) which expires and disperses energy and fluid, and secondarily an excessive condensation and concentration of the energy-fluid of the Liver sector (Mid-lower sector).	
Taeeum type Exterior Interior combined pattern with interior dominance	太陰人 裏病兼表證	A pattern characterized by interior symptoms that include systemic heat, ocular pain, dryness of nose, sleep disorder, tidal sweating, constipated bowel movement, thirst, etc; exterior symptoms that include cold-intolerance (accompanied with the absence of perspiration), dyspnea (labored breathing), somatic pain, diarrhea secondarily. It may be explained by: primarily the damage of the Esophagus sector (Frontal Upper sector) which expires and disperses energy and fluid, and secondarily an excessive condensation and concentration of the energy-fluid of the Liver sector (Mid-lower sector).	WHO-ICTM
Taeeum type Liver Heat pattern	太陰人 肝熱證	A pattern characterized by ocular pain, dryness of nose, sleep disorder, tidal sweating, constipated bowel movement, abdominal fullness, thirst, delirium. It may be explained by: An excessive condensation and concentration of the energy-fluid of the Liver sector (Mid-lower sector) causing the stagnation of energy and fluid, followed by the production of pathogenic heat.	WHO-ICTM
Taeeum type Liver Heat pattern accompanying Pestilential disease	太陰人 肝熱證 瘟病	A pattern characterized by cold-intolerance, systemic heat, somatic pain, and laryngopharyngeal swelling and pain in the Taeeum type Liver Heat pattern. It may be explained by: When signs and symptoms of plague-like infectious disease are present with an excessive condensation and concentration of the energy-fluid of the Liver sector (Mid-lower sector).	WHO-ICTM
Taeeum type Liver Heat pattern by Yang toxin	太陰人 陽毒證	A pattern characterized by facial erythema, swelling and pain in the laryngopharynx, and spitting of blood and pus in the Taeeum type Liver Heat	WHO-ICTM

Term	Chinese	Definition / Description	Ref
		pattern. It may be explained by: An accumulation of toxic heat (hyperthermia or hyperpyrexia) in the skin and flesh of the whole body caused by pathogenic heat with an excessive condensation and concentration of the energy-fluid of the Liver sector (Mid-lower sector).	
Taeeum type Liver Heat pattern with Yang Brightness symptoms	太陰人 陽明病證	A pattern characterized by somatic heat, absence of cold-intolerance, and constipated bowel movement in the Taeeum type Liver Heat pattern. It may be explained by: An accumulation of toxic heat (hyperthermia or hyperpyrexia) in the Small Intestine sector (Frontal Mid-lower sector)caused by pathogenic heat with an excessive condensation and concentration of the energy-fluid of the Liver sector (Mid-lower sector).	WHO-ICTM
Taeeum type Liver Heat pattern, NEC	太陰人 肝熱證, 他處未分類	A pattern characterized by ocular pain, dryness of nose, sleep disorder, tidal sweating, constipated bowel movement, abdominal fullness, thirst, delirium. It may be explained by the subpattern which is not else classified in the Taeeum type Liver Heat pattern.	WHO-ICTM
Taeeum type Supraspinal Exterior mild pattern	太陰人 背傾表病輕證	A pattern characterized by cold-intolerance (accompanied with the absence of perspiration), systemic heat, somatic pain, and dyspnea (labored breathing) in the Taeeum type Supraspinal Exterior pattern. It may be explained by: A decreased capbility of the Esophagus sector (Frontal Upper sector) to expire and to disperse energy and fluid, and insufficient immunological response to pathogens at the exterior level of the body.	WHO-ICTM
Taeeum type Supraspinal Exterior pattern	太陰人 背傾表病證	A pattern characterized by cold-intolerance (accompanied with the absence of perspiration), systemic heat, and dyspnea (labored breathing). It may be explained by: A decreased capbility of the Esophagus sector (Frontal Upper sector) to expire and to disperse energy and fluid.	WHO-ICTM

Term	Chinese	Definition / Description	Ref
Taeeum type Supraspinal Exterior pattern accompanying Cold reversal	太陰人 背顀表病寒厥證	A pattern characterized by alternating pattern of heat-based phases showing systemic heat without sweating and cold-based phases showing cold-intolerance without signs of systemic heat, which together make up its relative long-term course of disease in the Taeeum type Supraspinal Exterior pattern. It may be explained by: A decreased capbility of the Esophagus sector (Frontal Upper sector) to expire and to disperse energy and fluid, and the lingering state of an insufficient immunological responsiveness to pathogens at the exterior level of the body.	WHO-ICTM
Taeeum type Supraspinal Exterior pattern accompanying Wheezing and Panting.	太陰人 哮喘病證	A pattern characterized by wheezing and panting from dyspnea in the Taeeum type Supraspinal Exterior pattern. It may be explained by: A decreased capbility of the Esophagus sector (Frontal Upper sector) to expire and to disperse energy and fluid causing an imbalance in respiratory function.	WHO-ICTM
Taeeum type Supraspinal Exterior pattern, NEC	太陰人 背顀表病證, 他處未分類	A pattern characterized by cold-intolerance (accompanied with the absence of perspiration), systemic heat, and dyspnea (labored breathing). It may be explained by the subpattern which is not else classified in the Taeeum type Supraspinal Exterior pattern.	WHO-ICTM
Taeeumin; Taeeum type; TE type	太陰人	one who is given the organ structural scheme of a large Liver system and a small Lung system	WHO-IST
Taegeuk (Taiji); the Great Pole; the Supreme Ultimate	太極	in cosmogony, the transcendent state prior to differentiation into the Four Images; representative of the "Principle" (intrinsic, cause, principle)	
Taeyang type Exterior Interior combined pattern	太陽人 表裏兼病證	A pattern characterized by exterior symptoms that include systemic heat, cold-intolerance, somatic pain, and leg enervation (pseudo-paraparesis); interior symptoms that include abdominal pain, borborygmus, diarrhea, dysentery, dysphagia or regurgitation at a same time.	WHO-ICTM

Term	Chinese	Definition / Description	Ref
Taeyang type Exterior Interior combined pattern with exterior dominance	太陽人 表病兼裏證	A pattern characterized by exterior symptoms that include systemic heat, cold-intolerance, somatic pain, and leg enervation (pseudo-paraparesis); interior symptoms that include abdominal pain, borborygmus, diarrhea, dysentery, dysphagia or regurgitation secondarily. It may be explained by: primarily an excessiveness of the expirational dispersive energy of the Lung sector (Upper sector) and a weakening of the inspirational concentrative energy of the Liver sector (Mid-lower sector), and secondarily a strengthening in the expanding and dispersive force of the energy-fluid of the Esophagus sector (Frontal Upper sector) and a weakening in the condensing and concentrative force of the energy-fluid of the Small Intestine sector (Frontal Mid-lower sector).	WHO-ICTM
Taeyang type Exterior Interior combined pattern with interior dominance	太陽人 裏病兼表證	A pattern characterized by interior symptoms that include abdominal pain, borborygmus, diarrhea, dysentery, dysphagia or regurgitation; exterior symptoms that include systemic heat, cold-intolerance, somatic pain, and leg enervation (pseudo-paraparesis) secondarily. It may be explained by: primarily a strengthening in the expanding and dispersive force of the energy-fluid of the Esophagus sector (Frontal Upper sector) and a weakening in the condensing and concentrative force of the energy-fluid of the Small Intestine sector (Frontal Mid-lower sector), and secondarily an excessiveness of the expirational dispersive energy of the Lung sector (Upper sector) and a weakening of the inspirational concentrative energy of the Liver sector (Mid-lower sector).	WHO-ICTM
Taeyang type Exterior origin Lower back mild pattern	太陽人 外感腰脊病輕證	A pattern characterized by cold-intolerance, systemic heat, or somatic pain in the Taeyang type Exterior origin Lower back pattern. It may be	WHO-ICTM

Term	Chinese	Definition / Description	Ref
		explained by: An excessiveness of the expirational dispersive energy of the Lung sector (Upper sector), and a weakening of the inspirational concentrative energy of the Liver sector (Mid-lower sector), causing a condition of insufficient immunological responsiveness to pathogens at the exterior level of the body.	
Taeyang type Exterior origin Lower back pattern	太陽人 外感腰脊病證	A pattern characterized by systemic heat, cold-intolerance, somatic pain in the mild stage and leg enervation (pseudo-paraparesis) in the advanced stage. It may be explained by: An excessiveness of the expirational dispersive energy of the Lung sector (Upper sector) and a weakening of the inspirational concentrative energy of the Liver sector (Mid-lower sector), which damages the lower back area (dorsal Mid-lower sector) that is associated with the Liver Sector.	WHO-ICTM
Taeyang type Exterior origin Lower back pattern accompanying Leg enervation	太陽人 解㑊證	A pattern characterized by difficulty in ambulation due to enervation of the Lower body while the upper body remain sintact (the symptoms of cold-intolerance, systemic heat, or somatic pain are absent) in the Taeyang type Exterior origin Lower back pattern. It may be explained by: An excessiveness of the expirational dispersive energy of the Lung sector (Upper sector) and a weakening of the inspirational concentrative energy of the Liver sector (Mid-lower sector), producing upper excess and lower deficiency, which damages the lower back area that is associated with the Liver Sector.	WHO-ICTM
Taeyang type Exterior origin Lower back pattern, NEC	太陽人 外感腰脊病證, 他處未分類	A pattern characterized by systemic heat, cold-intolerance, somatic pain in the mild stage and leg enervation (pseudo-paraparesis) in the advanced stage. It may be explained by the subpattern which is not else classified in the Taeyang type Exterior origin Lower	WHO-ICTM

Term	Chinese	Definition / Description	Ref
		back pattern.	
Taeyang type Interior origin Small Intestine mild pattern	太陽人 內觸小腸病輕證	A pattern characterized by abdominal pain, borborigmus, diarrhea, or dysentery in the Taeyang type Interior origin Small Intestine pattern. It may be explained by: A decrease in the function of energy-fluid metabolism in the Small Intestines (Frontal Mid-lower sector) with a strengthening in the expanding and dispersive force of the energy-fluid of the Esophagus sector (Frontal Upper sector) and with a weakening in the condensing and concentrative force of the energy-fluid of the Small Intestine sector (Frontal Mid-lower sector).	WHO-ICTM
Taeyang type Interior origin Small Intestine pattern	太陽人 內觸小腸病證	A pattern characterized by abdominal pain, borborygmus, diarrhea, dysentery in the mild stage and dysphagia or regurgitation in the advanced stage. It may be explained by: A strengthening in the expanding and dispersive force of the energy-fluid of the Esophagus sector (Frontal Upper sector) and a weakening in the condensing and concentrative force of the energy-fluid of the Small Intestine sector (Frontal Mid-lower sector), producing the imbalance of the energy and fluid metabolism and subsequent damage of energy and fluid.	WHO-ICTM
Taeyang type Interior origin Small Intestine pattern accompanying Regurgitation	太陽人 噎膈證	A pattern characterized by inability to accept ingested food and regurgitation of the food stuff (the symptoms of abdominal pain, borborigmus, diarrhea, or dysentery are absent) in the Taeyang type Interior origin Small Intestine pattern. It may be explained by: The severe damage of the energy-fluid in Esophagus sector (Frontal Upper sector) with a strengthening in the expanding and dispersive force of the energy-fluid of the Esophagus sector (Frontal Upper sector) and with a weakening in the condensing and concentrative force of the energy-fluid of the Small Intestine sector (Frontal Mid-lower sector).	WHO-ICTM

Term	Chinese	Definition / Description	Ref
Taeyang type Interior origin Small Intestine pattern, NEC	太陽人 內觸小腸病證, 他處未分類	A pattern characterized by abdominal pain, borborygmus, diarrhea, dysentery in the mild stage and dysphagia or regurgitation in the advanced stage. It may be explained by the subpattern which is not else classified in the Taeyang type Interior origin Small Intestine pattern.	WHO-ICTM
Taeyangin; Taeyang type; TY type	太陽人	one who is given the organ structural scheme of a large Lung system and a small Liver system	WHO-IST
the avaricious man	貪人	one who discards benevolence and succumbs to acquisitiveness	
the frivolous man	薄人	one who discards wisdom and succumbs to pretentiousness	
the ignoble man	鄙人	one who discards propriety and succumbs to indiscipline	
the Path and Virtue	道德	the destination point of accumulated "Knowledge" and "Action"	
the timid man	懦人	one who discards righteousness and succumbs to stealthiness	
thoracic area	膈	the site related to the stomach; the anterior body part related to the mid-upper burn-center	
to admire the wise and appreciate the noble	好賢樂善	to admire the wise one and appreciate the noble one; the cure for disease	
to envy the wise and covet the able	妬賢嫉能	to envy the wise one and to covet the able one; the cause of disease	
to scorn evil	惡惡	the basis for doing the Acting that can lead to the Proper Course	
to value good	好善	the basis for doing the Knowing that can lead to the Ultimate Nature	
Ui-Ryeo-Jo-Ji	意慮操志	the substance stored in the Four Anterior Pools, related to mental and psychological functioning	
umbilical area	臍	the site related to the small intestine; the anterior body part related to the mid-lower burn-center	
understanding of the others and rectification of the self	知人正己	the method of proper knowing and acting through rejection of the inattentive mind (by preserving the mind and cultivating the nature) and rejection of the indolent conduct (by rectifying the body and propping the	

Term	Chinese	Definition / Description	Ref
		course)	
upper back and neck	背傾	upper spine; a site pertinent to the lung	
upper back area	膂	the site related to the spleen; the posterior body part related to the mid-upper burn-center	
upper burn-center; uppermost burn-center	上焦	one of the Four Burn-Centers, the sector related to the lung and esophagus, extends to the concept of the Lung Affiliation	
upper spine	背膂	the site where the clear energy of the pool of Go, after coming out by the driving force of the eye (force driven by seeing) to be thence converted into Gi, goes in to form the pool of Mak ; the body part pertinent to the mid-upper burn-center	
urgency	急迫之心	constant urgency; the general mindset of the Taeyang type	
vast-flowing energy	浩然之氣	the result of expansion and fulfillment of the energy of the Lung, Spleen, Liver, and Kidney (benevolence, righteousness, propriety, and wisdom)	
vast-flowing principle	浩然之理	the result of illumination and discrimination of the greed of the Heart (ignobility, frivolity, avarice, and timidity)	
vigilance (alertness)	警戒	one of the tonic factors, avoidance of authority	
waist	腰圍	the body part pertinent to the mid-lower burn-center	
warm-hot-cool-cold energies	溫熱凉寒	the four energies derived from water-food; energies formed when the energy derived from water-food is separated within the Four Bu Organs according to its clarity or turbidity by the mechanism of ascent and descent in the superior-inferior dimension	
worldly (social) interactions	世會	one of the Four Heavenly Frames; the form of society or the characteristics of the nation that are forged as social relationships become more diverse (society-related element)	
yang-warm qi	陽煖之氣	The qi that promotes health for Soeum type.	WHO-IST
yin-cool qi	陰淸之氣	The qi that promotes health for Soyang type.	WHO-IST

저 자

이의주
한의학박사, Medical MBA, 교수
경희대학교 한의과대학 사상체질과
교환교수, 메릴랜드의과대학 CIM

이지원
한의학석사, 전문의(사상체질과)

유정희
한의학박사, 간호학석사
박사후과정, 메릴랜드의과대학 & 존스홉킨스대학

한국학연구총서 48
사상의학 값 28,000원

2014년 3월 30일 1판 1쇄

저 자 이의주 · 이지원 · 유정희
발 행 인 임 삼 규
발 행 처 **지 문 당**
주 소 10881 경기도 파주시 광인사길 85(본사)
03134 서울시 종로구 돈화문로 82(집문당 5층)
등 록 1997. 12. 30. 제406-2003-000038호
영 업 부 (02)743-3192~3 팩스(02)742-4657
전자우편 sale@jimoon.co.kr
편 집 부 (02)743-3096~7 팩스(02)743-0227
전자우편 edit@jimoon.co.kr
홈페이지 www.jimoon.co.kr

ISBN 978-89-6297-152-1
978-89-6297-002-9(세트)